Robert Jackson

**A Treatise on the Fevers of Jamaica**

With some observations on the intermitting fever of America, and an appendix,

containing some hints on the means of preserving the health of soldiers in hot

climates

Robert Jackson

**A Treatise on the Fevers of Jamaica**
*With some observations on the intermitting fever of America, and an appendix, containing
some hints on the means of preserving the health of soldiers in hot climates*

ISBN/EAN: 9783337308896

Printed in Europe, USA, Canada, Australia, Japan

Cover: Foto ©berggeist007 / pixelio.de

More available books at **www.hansebooks.com**

A

# TREATISE

ON THE

## FEVERS OF JAMAICA,

WITH SOME

*OBSERVATIONS*

ON THE

## Intermitting Fever of America,

AND AN

# APPENDIX,

CONTAINING SOME HINTS ON THE MEANS OF PRE-
SERVING THE HEALTH OF SOLDIERS IN HOT
CLIMATES.

By ROBERT JACKSON, M.D.

*Ne mea dona tibi studio disposta fideli,*
*Intellecta prius quam sint, contempta relinquas.* Lucret. lib. I.

—Διατειρα τοι
Βροτων ελεγχος. Pind: Olym. 4.

*PHILADELPHIA:*
PRINTED FOR ROBERT CAMPBELL, BOOKSELLER.

1795.

THE obfervations contained in the following pages, were made during the time that I lived in Jamaica, or while I attended fome part of the army in America. The materials were collected between the years 1774 and 1782; and the prefent performance would have been offered to the public before this time, had I fooner found leifure to attend to the bufinefs of publication. The infufficiency of Dr. Hillary's work, the moft efteemed book on the difeafes of the Weft Indies, and the only one with which I was acquainted while I remained in that country, furnifhed me with a motive for the undertaking ; a motive, which may be thought, perhaps, no longer to exift, as two treatifes have been publifhed lately by Dr. Hunter and Dr. Mofcley, exprefsly on the difeafes of which I have written. I might remark, that the prefent attempt, fuch as it is, was nearly completed before the trea-tifes to which I allude came to my hands. The views which they afford of fevers, as they differ from each other, fo they likewife differ from thofe which I have ventured to advance. I have weighed their merits maturely, and cannot fay that I difcover any information, which gives me caufe to change thofe opinions which I had formed, or which renders the publication of the prefent work unneceffary. I may obferve, that Dr. Hunter details, with candour and per-fpicuity, the mode of practice, which was followed by the moft refpectable medical people of Jamaica, at the time that I lived in the ifland. He perhaps prefcribes the bark in larger quantities, than was then cuftomary; but I do not perceive any thing in the plan of treat-ment effentially new; neither will Dr. Mofeley, though he purfues innovation with great eagernefs, be better able to eftablifh his claim to original difcove-ries. The plentiful and long continued purging, on which he places a confider-able fhare of his merit, has been a fa-vourite practice with numbers for many years paft; and the free ufe of the lancet, which he recommends fo much in fevers, was employed in feveral diftricts of

Jamaica, before this author's name was known. Dr. Spence, a practitioner of fome eminence at Lucea, in the Weftern extremity of the ifland, wrote a pamphlet (I believe in the year 1776) with a view to enforce its fafety and utility, in promoting the cure of the general clafs of febrile difeafes. The publication was well received, and ferved to remove the dread of the lancet, which fome people till then had falfely entertained.

I have thus explained the motive which induced me to write ; and I leave it to people of experience to judge of the manner in which I have executed the tafk. I fhall only obferve, that, when I firft intended to prepare my obfervations for the infpection of the public, I had no other defign than to mention

fuch circumftances in the hiftory and
cure of fevers, as did not appear to be
generally known. Having fpent the
earlier part of my life in fituations,
which did not admit of a continued plan
of ftudy, I had not till lately much acquaintance with the opinions of medical
writers. About four years ago, when
I found a fettled abode, I began occafionally to look into the works of the
ancient phyficians. In examining Hippocrates, I difcovered fuch a fimilarity
in the fevers of the Archipelago, with
thofe of Jamaica, that I fufpended my
defign of publifhing, till I fhould have
given that author's writings a careful
perufal. I foon was convinced that
many obfervations, which I had confidered as my own, were actually known
to this father of phyfic ; and though I

was probably difappointed in being anticipated in offering fomething new to
the public, I was ftill gratified by the
coincidence of remark, that gave me a
confidence in my accuracy which otherwife I durft not have affumed. After
I had carefully perufed the writings of
Hippocrates, I confulted and compared
fuch other of the Greek phyficians, as I
was able to procure. I even defcended
with a fimilar examination to the prefent times ; but as my collection of medical writers, particularly · of modern
ones, is fmall, I have probably omitted
fome who ought to have been mentioned, or perhaps advanced obfervations as my own, which in reality belong to others. If I have done fo, I
muft be allowed to fay, that I have done
it without confcioufnefs.

It may appear, perhaps, that I have treated the opinions of great names with too little refpect; but if facts have at any time occurred to me, which contradict eftablifhed theories, I fhould hope that thefe facts will be examined before they are rejected. No medical authority ought to prevail over the certain evidence of truth. I am not confcious of having mifreprefented, for the fake of a theory, the minuteft circumftance of what I have actually feen ;— if I have been miftaken in any inftance in forming conclufions, I fhall receive the correction of my inaccuracies with gratitude.

# CHAP. I.

A S it is perfectly well known, that fevers, which
are effentially in themfelves the fame difeafe,
vary in their appearances, from difference of climate
and feafon, it would be very fuperfluous to attempt
any proof of what is fo generally acknowledged. Even
Hippocrates, who lived more than two thoufand years
ago, does not feem to have·been unacquainted with
this fact. This induftrious phyfician, as we learn
from many paffages in his works, not only vifited the
various iflands in the Ægean Sea; but travelled likewife into various parts of the adjacent continents. The
motive of his journeys, we are taught to believe, was
principally to obferve the different fituation of places,
and to mark their correfponding difeafes. He has
defcribed very fully, in a curious and ufeful treatife,
the effects of air and local fituation on the human
frame; and in the hiftory of fevers, related in the books
of Epidemics, has feldom forgotten to mention, not
only the general conftitution of the feafons with refpect to difeafes; but likewife the nature of the climate, where his obfervations were more particularly
made. The fpecies of fevers, it is true, that are found
in the writings of Hippocrates are extremely multiplied; yet this author feems ftill to have been clearly
of opinion, that difeafes, which are effentially the fame,
affume, in fome refpects, a different appearance in the
ifland of Thafus, and at Abdera, on the contiguous

**B**

coaft of Thrace. Obfervations to the fame effect
have been made by many others in different parts of
the world; nor is the fact capable of being better illuftrated in any country than in Jamaica; where a
very fmall change of place, or the ordinary revolution
of feafons, are often obferved to influence in a high
degree, the appearances of the common endemic of
that ifland. But this obfervation,—that local fituation and the change of feafons, are the caufe of varieties in the appearances of the fame difeafe, has been
fo long known, that it is almoft unneceffary to mention it; nor would it have been repeated now; unlefs
to obviate any objections which might be made to the
hiftory, that is given in the following pages, fhould it
not be found exactly to correfpond, with that which
has been obferved by others, who have lived in the
other iflands of the Weft Indies, or in other parts of
Jamaica; or perhaps even in the fame part of Jamaica,
at a different period of time. It is prefumed, indeed,
that the fame fundamental diftinctions of fever obtain
in every ifland within the tropics; yet the defcription
that is given in this place, (it may not be unneceffary
to mention), is affirmed to be ftrictly exact, only in
the diftrict about Savanna la Mar; and that only for
a fhort fpace of time: viz. from the year 1774 to the
year 1778.

The fever, that chiefly prevailed at Savanna la
Mar during the period mentioned above, was ufually
mild in its fymptoms, and more regularly remitting
in its form than I underftood the endemic difeafe to
be in moft other parts of the ifland. To what circumftances in the local fituation this might be owing,
I do not pretend to determine. Savanna la Mar is
fituated clofe by the fea: its particular fcite, and the
furrounding country to the diftance of feveral miles
is perfectly level; whilft in confequence of its being
open to the fea on the eaft, it is vifited early, and conftantly by a falutary and refrefhing breeze. There

is a fmall rivulet, indeed, which, lofing itfelf almoft
entirely in mud, forms a morafs that partly furrounds
it on the north. From vicinity to this morafs the fituation of Savanna la Mar has been fufpected to be unhealthy; but it is not fo in fact. The fea at high water, particularly in the fpring tides, overflowing the
fwampy ground, contributes in a great meafure, pre-
haps, to diminifh the more ufual noxious qualities of
the marfh exhalation. The opinion of many eminent
writers, however, is not altogether favourable to this
idea; but there feems to be reafon to doubt, whether
the opinion formed by thofe writers in this inftance,
is the refult of accurate and careful obfervation; or if
it is merely a fuggeftion of theory. I am inclined to
the latter way of thinking, as I have never found the
neighbourhood of Salt-marfhes, in the different parts
of America that I have had the opportunity of vifiting,
lefs healthful than the reft of the country:—on the
contrary, they were frequently more fo.

Before proceeding to a particular defcription of the
fever, that chiefly prevailed in the diftrict of Savanna
la Mar, it may not be fuperfluous to give the outline
of its character, that we may be the better enabled to
trace its refemblance with the fevers of other climates,
and to determine its place in the general clafs of febrile difeafes. The common fever of Savanna la Mar,
which, as I mentioned before, was ufually mild in its
fymptoms, and regular in its form, feems to be
properly ranked with thofe, that ftrictly fpeaking
are called remitting. It is difficult to fix the boun-
daries between remittents and intermittents, if a fixed
boundary actually exifts. The paroxyfms of the fever
of Jamaica are obferved, in many inftances, to terminate in more perfect remiffions than the paroxyfms
of the endemic of North America, which is known to
be fundamentally an intermitting difeafe. Hence authors generally are of opinion, that all the difference
which appears to take place in thofe fevers, depends

merely on an accidental cause: viz. on the greater or
less heat of the climate. I shall not be positive that it
is not so; yet I cannot help remarking that certain
appearances incline me to be of opinion, that there
subsists, between the endemic of Jamaica and the en‐
demic of North America, a difference, in some degree,
fixed and essential. I cannot pretend to ascertain the
difference precisely; or to offer a conjecture about
the modification of the morbid cause in which it con‐
sists; yet the following circumstances give reason to,
believe, that to a certain degree at least it actually
takes place. The common fever of Jamaica, for in‐
stance, was not only disposed to terminate of its own
accord; but it was disposed to terminate on certain
critical days, often at an early period, and by signs of
crisis too clear to be mistaken; neither did the Peru‐
vian bark, in the manner at least in which it was ma‐
naged, ever cut short its course with certainty. The
endemic of America on the contrary often lasted long.
It frequently, indeed, changed to another disease after
a length of time; but no period could be assigned for
its natural termination. The signs of crisis, it may
likewise be remarked were so obscure as scarcely to
be distinguished with the closest attention; at least for
my own part I will own, that after an experience of
several years, and the greatest care in observing the
minutest circumstances, I never yet was able to say
with confidence, that the endemic of America, parti‐
cularly in the northern provinces, was gone not to re‐
turn again, till the hour of its return was past: nei‐
ther did the Peruvian bark, though its effects were
so equivocal in the fever of Jamaica, scarcely ever fail
of stopping the progress of this disease. To which
we may add, that the complaint, which strictly
speaking is called the intermittent, or ague and
fever, can scarcely be said to belong to Jamaica; at
least it was not known at Savanna la Mar. In the
course of four years I did not once observe it; and
those, who had lived much longer at this place, assured

me they had never feen an inftance of it; unlefs in perfons who were newly arrived from aguifh countries. This is a fact of fome importance, as it fhews to us, that though the proper intermittent is not the endemic difeafe of the country it is ftill capable of exifting in the climate:—no weak argument, that, the two difeafes of which we have been fpeaking, actually do poffefs fome difference in the modification of the general caufe, though we are unable to define the precife bounds and limits of it.

The circumftances which I have mentioned might incline us to be of opinion, that the fever of Jamaica, and the common endemic of America are not exactly the fame difeafe; that is, that though they depend on the fame general caufe; yet that this caufe undergoes fome fixed and permanent modification in thofe different climates, fo that an effential difference actually arifes. But though this really appears to be the cafe; yet I muft acknowledge, that the reigning epidemic of the fouthern provinces of America, often lofes its diftinctive marks of intermiffion in the hot months of fummer; whilft it approaches, in other refpects, fo near to the fever of Jamaica, as to be diftinguifhed from it with difficulty. This was particularly the cafe at Ebenezer in Georgia, in the year 1779, during the months of June and July. A cold fit was feldom obferved in this place; unlefs perhaps in the firft attack; lownefs, languor, head-ach, pain of the back and other difagreeable feelings remained even in the moft perfect remiffions: the difeafe was likewife much difpofed to terminate of its own accord: on the ufual critical days, frequently by figns of crifis, that were far from being obfcure: yet though thefe refemblances were fo very ftriking, the fever of Ebenezer was perfectly under the controul of Peruvian bark, which was not exactly the cafe with that of Jamaica. I muft remark, however, that bark was ufed with a freedom in America, that I never thought of attempting in the Weft Indies.

But though it ftill may be thought doubtful by fome, whether the autumnal fever of aguifh countries, and the endemic fever of Jamaica are characteriftically different, or effentially the fame difeafe; yet it is not fo difficult to trace a fpecific refemblance, between this fever of the Weft Indies, and the prevailing endemic of the Ægean Sea. The fever defcribed in the Epidemics of Hippocrates has every ftriking feature of the difeafe which is the fubject of the following treatife. The general character is the fame ; the courfe and mode of termination are often alike. If the duration is fometimes different, it is probably much owing to the more decifive practice of the moderns: for though it remains uncertain, whether a remedy has yet been difcovered, which abfolutely cuts fhort the fever of Jamaica in the midft of its courfe ; yet no doubt remains, that there are various modes of treatment, which may, and actually do render the ordinary changes of the critical day, decidedly critical. The fevers of the Ægean Sea, as defcribed by Hippocrates, and of Minorca, as defcribed by the accurate Cleghorn, bear the neareft refemblance to the endemic of Savanna la Mar. The fevers of Italy, of different parts of the continent of Afia, as defcribed by various writers, as well as the fever of America, of which I have had perfonal experience, however obfcure their remiffions, feem rather to be degenerated intermittents, than the difeafe defcribed in the following pages. But that I may not be thought to infift too much on this opinion, which few, perhaps, may confider as of much confequence, I fhall relate the hiftory of the fever of Savanna la Mar, as accurately as I can, leaving it to others to determine, whether it fhould be confidered as an intermittent degenerated into a remitting form, in confequence of the greater heat of the climate, or allowed to poffefs fomething characteriftic in its own nature.

# CHAP. II.

A Knowledge of the types and periods of fevers,
though fo neceſſary ,to be well underſtood by
thoſe who pretend to cure the diſeaſe, has unfortunately been little attended to by the practitioners
of Jamaica. In that country, I met with ſome who
were able to foretel the returns of the ſingle tertian;
but I met with none, and I have good reaſon to believe that there actually were few, who troubled themſelves about forms of greater complication. There
were many, on the contrary, who, blinded by theories
of accumulated bile, ridiculed altogether the idea of
this ſtated regularity in the movements of nature.
To enter into a diſpute with ſuch would be labour
loſt. The exiſtence of a regular type in the fevers
of the Weſt Indies does not admit of a doubt, yet we
muſt not forget to mention, that thoſe types are traced
with greater difficulty in that country, where remiſſions
are obſcure, than in others, where every paroxyſm is
uſhered in by a cold fit. In Jamaica it is impoſſible
to follow them without the moſt careful attention, or
perhaps to attain a clear and ſyſtematic knowledge of
them, without writing down and analyzing many of
thoſe caſes which occur in practice. I remember to
have been impreſſed with the idea, at a very early period, that one obſervation made by myſelf would in
reality be more uſeful than twenty equally important
in themſelves, which I only retained in my memory
from reading. Convinced of this truth I ceaſed to
look for information in books, from the time that I
arrived in the Weſt Indies ; but, truſting wholly to

my own experience, wrote down minutely, in the prefence of the fick, the hiftory and cure of the moft important cafes of fever that occured to me. At ftated
times I reviewed that which I had done, and arranged
under proper heads the moft ftriking circumftances,
that I found recorded in my notes. Among other unexpected appearances I obferved a regularity and order
in the types of fevers, of which at firft I had no idea.
But though the laws of nature appeared to be fixed
and ftable, in producing this varied but ftated regularity of form; yet a knowledge of thofe laws was not
eafily attained. Two years were fpent, and not fewer
than a hundred cafes were analyzed, before my views
of the fubject were in any degree accurate. The labour, perhaps, was fuperfluous; there being many authors who have defcribed minutely every variety and
every combination of type that has ever been obferved
to take place. But as I had little knowledge of the
writings of others at this period, I fhall content myfelf in the firft place with relating the hiftory of types
as they occurred to my own obfervation; at the fame
time that I fhall not omit to take notice occafionally
of the more conftant peculiarities, which are found
in authors of credit, who have practifed in different
countries. The influence of climate I may obferve is
of confiderable effect in modifying the various forms.

'The fingle tertian, the period of which is forty-
eight hours, is a form of fever that occurred frequently
in Jamaica, particularly in the dry and healthy feafon.
Its courfe was eafily traced, as the remiffions were
often diftinct, and the acceffions fometimes diftinguifhed by a flight horror or fhivering.

But though the above-mentioned type was by no
means of rare occurrence at Savanna la Mar; yet the
double tertian, with fimilar paroxyfms on alternate
days, was ftill more common, particularly in the rainy
and fickly months. This form of fever, which feemed
to confift of two fingle tertians, that ran a feparate

and independent courfe, began ufually in the morning.
Its hour of invafion was from eight to ten; and its
acceffion was generally diftinguifhed by a cold fit.
The paroxyfm, which for the moft part was regularly
formed, declined after a continuance of eight or ten
hours; and the patient remained free from fever, not
only during the night, but during the following day,
till four in the afternoon, or later. A feverifh indifpofition ufually came on then, which continued the
whole or the greateft part of the night. A remiffion
took place; but it was foon fucceeded by a paroxyfm
fimilar in its fymptoms, and manner of attack, to the
paroxyfm of the firft day. This having declined towards evening, the patient, as formerly, was free from
fever during the night and the day following. At the
ufual hour, however, or more generally before it, a
paroxyfm more diftinctly formed in all its parts, and
more violent in degree, than the preceding one which
correfponded with it, returned, and continued till
morning. It fubfided; and was foon fucceeded by the
paroxyfm of the fifth day, which declined, as formerly,
after the ufual duration. Such was the general courfe
and progrefs of the difeafe.—The fever of the odd day,
which began for the moft part in the morning, ufually
returned later and later, and with decreafing violence
every fucceeding paroxyfm; whilft the fever of the
even day, which generally began in the afternoon or
evening, as ufually returned earlier, and when this
was the cafe, frequently encreafed in force. The fever which came on in the morning generally began
the difeafe. It was for the moft part a fever of complete and regular paroxyfms from the firft attack.
The fever of the evening, on the contrary, was little
more than an indifpofition in its beginning; nor was
its time of appearing at all certain. It often was not
perceived till the evening of the fourth day; fometimes not till later, neither was its continuance of a
fixed duration. Sometimes it went on after the fever

of the morning had ceafed; and in other cafes it terminated, while the other purfued an uninterrupted courfe.

The type, which was fo frequent in Thafus, and the other iflands of the Ægean Sea, feems to be a fpecies of the double tertian. Mention is made frequently of fuch a form in the Epidemics of Hippocrates; yet the double tertian of Hippocrates is directly oppofite, in fome refpects, to the type which I have juft now defcribed.— The evening fever generally began the difeafe in thofe iflands: hence the great exacerbations, and confequently the crifis were frequently on the even days. Such alfo appears to have been fometimes the cafe in Minorca: yet Cleghorn acknowledges, that a type, fimilar to that which I have defcribed as prevailing fo commonly at Savanna la Mar, was by much the moft frequent form of the abovementioned ifland. It was alfo, I may add, frequent in the fouthern provinces of North America, particularly in the hot months of fummer and autumn.

A quotidian type has been defcribed by almoft every author, who has written on the fubject of intermitting and remitting fevers: neither can it be denied, that forms of difeafe frequently occur, where the paroxyfm returns every day at the fame hour, with fymptoms fo nearly alike, that an ordinary obferver can perceive no difference. Yet Mr. Senac, an author of confiderable eminence, boldly maintains, that a real quotidian type does not exift in nature. The queftion, perhaps, is not eafily determined. I can only mention from my own experience, that I have frequently feen fevers with quotidian exacerbations of fuch a kind, that there was no perceivable difference between them, either in fymptoms or in manner of attack; yet I muft likewife own, that thefe exacerbations were generally in the evenings; and that the difeafe was not in every refpect of a diftinct intermitting form.

I have now mentioned thofe types that are moft

frequently met with in the fevers of Jamaica; yet befides thefe, there now and then occur others of more
complicated and perplexing forms. Thus, I have
fometimes feen at Savanna la Mar, a fever, which
feemed to be compounded of a fingle tertian with a
quotidian. In the fpace of forty-eight hours there
were three feparate exacerbations, two of which were
exactly alike, that it was impoffible to perceive a difference; whilft the whole three followed each other
in fuch an order of time, that to a fuperficial obferver, there appeared to be only one long paroxyfm of
thirty-fix hours or more. An example will make
it more plain. On Monday, for inftance, the paroxyfm of a fever was obferved to begin about nine
in the morning, preceded by fome degree of coldnefs
and fhivering. This paroxyfm was ufually violent
in its firft attack: But declined gradually towards
evening; though before it was gone off totally,
another exacerbation commenced, which continued
the whole of the night. This likewife abated on
Tuefday morning; and the patient remained apparently free from fever till five in the afternoon or
later. A paroxyfm then made its appearance, fimilar
to the paroxyfm of the preceding evening. It ran
over a courfe of fimilar duration, and fcarcely had
abated on Wednefday morning, when a paroxyfm refembling that of Monday fucceeded it; which, as formerly, declining towards evening, was foon followed
by an exacerbation, that lafted till the morning of
Thurfday. On Thurfday, as on Tuefday, there was
no fever till late in the afternoon; when the evening
exacerbation returning at the ufual hour, proceeded
in its ufual courfe.

The tertian type, fimple, or varioufly compounded,
is the form of fever, which prevails moft univerfally
in all climates. The quartan in thofe countries were
I have lived was rare; and the exiftence of a real quotidian, perhaps, is doubtful. To that compound form,

which I have juſt now deſcribed, I ſhould be inclined
to give the name of Semitertian. It is exprefsly the
difeafe, which I now find has been deſcribed under
this name by Hoffman ; but it is more difficult to determine exactly, if it is the Hemitritæus of the ancients. The Hemitritæan form is frequently mentioned by Hippocrates; but his definition is too lo ſe
to enable us to judge precifely of its nature. It is in
fact impoſſible to ſay with certainty, whether it is to
a type ſimilar to that which I have deſcribed, or to the
extended and ſubintrant paroxyfms of the double tertian, that he has applied the name. This laſt, indeed,
ſeems to have been the idea of many of the ancients,
particularly of Celſus and Agathinus.—The phyſicians of the earlier ages, were leſs curious in minute
diſtinctions than their followers ; and probably applied
the name of Hemitritæus to thoſe fevers, the paroxyfms of which were ſimply of unuſual duration, no
leſs than to thoſe that were of a complicated or compound nature. This actually appears to have been
the caſe for a great length of time; but at laſt, and
not long indeed before the days of Galen, the ſophiſ-
tical genius of the profeſſors of medicine, which exerted itſelf chiefly in things of little importance, multiplied the types of fevers to an endleſs variety, and
attempted to eſtabliſh diſtinctions which have no exiſtence in reality. Galen, who is copious in moſt
things, has difcuſſed very fully the ſubject of types in
general, and deſcribed particularly the nature of the
Semitertian at great length. But whatever credit
may be otherwiſe due to the obſervations of this il-
luſtrious writer, it is evident that the deſcription, in
the preſent inſtance, is merely the reſult of theory.
He has attempted, indeed, to illuſtrate his opinion by
an example ; but the caſe he has furniſhed us with,
is conſtantly varying its hour of attack ; and if accurately examined, appears rather to be a triple
tertian, properly ſo called, than the form of fever that

I have defcribed above. After the Greeks we may next take a fhort view of the Arabian phyficians, who, as they borrowed much of their knowledge from the writings of Galen, likewife adopted his idea of the compound nature of the Semitertian. Avicenna, the moft eminent among them, has defined this form of difeafe with a good deal of precifion; but we do not find that he has added any thing very material to the opinions of his predeceffors. Galen indeed, had difcuffed the fubject fo fully, that, though fucceeding writers fometimes changed names, they do not feem in reality to have furnifhed much new obfervation. As we defcend to lefs remote times, Hoffman and Cleghorn are the moft accurate of the moderns, on this fubject, who have yet come to my hands. The former defcribes under the name of Semitertian the exprefs form of difeafe, that I met with in Jamaica; the latter refers this title to the extended and fubintrant paroxyfms of the double tertian. It is with unwillingnefs that I differ in opinion from Cleghorn, who has thrown more light on the hiftory of periodical fevers, than perhaps, all the writers of his time : yet I cannot help obferving, that I never recollect to have met with an original Semitertian, fuch as that he has defcribed in his well-known treatife. I muft own, indeed, that I have feveral times feen the morning fever of the double tertian anticipate, fo as to be mixed with the decline of the paroxyfm of the preceding evening; thereby producing a form of difeafe, that could only be diftinguifhed, by the moft careful attention, from one long paroxyfm of thirty hours or more : yet this was in fact, only a degenerated double tertian, the paroxyfms of which became accidentally mixed with each other.

The types which I have defcribed above are all the varieties, that I obferved in the fevers of the Weft Indies; yet more extenfive experience might have, perhaps, brought to my knowledge ftill further com-

C

plications : for befides the forms mentioned in the preceding pages, I had the opportunity, in the fouthern
provinces of North America, to fee feveral inftances
of the triple tertian, properly fo called. In forty-
eight hours there were three feparate paroxyfms ; all
of them differing from each other, but correfponding
with others that followed.—They were eafily traced,
as the acceffions in that climate were ufually diftinguifhed by a cold fit. In Jamaica, on the contrary,
complications and irregularities were marked with
difficulty. The anticipation of type among other
things occafioned confiderable perplexity. Thus the
fingle tertian, whofe regular period is forty-eight
hours, often completed its revolutions in forty-fix.
But though the paroxyfms frequently returned one
hour or even two hours fooner than the ufual time;
yet thefe anticipations were feldom longer, unlefs the
difeafe was of a malignant nature, or difpofed to change
to a continued form. In either of thefe cafes anticipations of twelve or fourteen hours were not uncommon. Of the different forms of fever the anticipations of the fingle tertian were the longeft and moft
remarkable. It was obferved, indeed, that the evening paroxyfm of the double tertian very generally returned before the ufual hour ; yet this return was feldom more than an hour at once ; nor did the time of
invafion in the whole courfe of the difeafe, fo far as I
have obferved, ever go beyond twelve at noon. The
morning paroxyfm, on the contrary, frequently poftponed; yet I have likewife obferved it fometimes to
anticipate fix or eight hours at one time; by which
means, it encroached on the paroxyfm of the preceding
evening, and produced the difeafe which Cleghorn has
diftinguifhed by the name of Semitertian. The obfervation of this fact does not feem to have efcaped
Avicenna.

There undoubtedly are accidental circumftances,
which have fome effect in accelerating or retarding

the return of the paroxyfm for a fhort fpace of time;
yet it would appear, upon the whole, that the caufe,
why a fever anticipates or poftpones, depends upon
fomething that is peculiar to the particular nature of
the difeafe. Thus, a fever, which once begins to anticipate, generally goes on anticipating through a
great part of its courfe; a certain proportion being
frequently preferved between the anticipations of the
different paroxyfms. This was particularly the cafe
in the fingle tertian of America. The paroxyfms of
the fevers of that country often anticipated to a certain point, by fhort anticipations, which bore a regular proportion to each other; whilft they were fometimes likewife obferved to poftpone, in the fame gradual manner, towards the termination. This feldom
happened in the fevers of the Weft Indies.

The anticipations I have mentioned often occafion
perplexity in tracing the types of fevers; yet the complications which arife in the progrefs of the difeafe,
have a ftill more confiderable effect in embarrafling
the appearances of regularity. Fevers which begin
as fingle tertian, often continue fuch through the
whole of their courfe; yet it fometimes likewife happens, that complicating fevers make their appearance
on the even days for inftance, and continue longer,
or terminate fooner than the original complaint, in
fuch manner, as if their exiftence no way depended
on it. To be able to diftinguifh thofe complications
from the anticipations of a fingle type is frequently
very ufeful, and a knowledge of it may, in general, be
attained with a good deal of certainty. Thus for
inftance, if the difeafe is moderate in its fymptoms, and
without fufpicion of malignity, the appearance of a
paroxyfm twelve or fourteen hours before the ufual
time, efpecially if there is no material difference in
the nature of the fymptoms, gives reafon to fufpect
that the premature return is in reality the complication
of another fever. On the contrary, where the difeafe

C 2

has betrayed figns of malignity, or where the fymptoms differ from thofe of the former paroxyfms only
in a greater degree of violence, there is then reafon
to apprehend that this appearance is only an anticipation. On the fubject of complication of type, fome
curious obfervations may be found in the writings of
Galén and Avicenna. The opinions of thofe authors, indeed, are often mixed with whimfical theories; yet in many refpects they are fundamentally
true and highly interefting.

It would be a matter of fome utility could we learn
to foretel, from the nature of the types, the accidents
that are likely to happen in the progrefs of the difeafe,
or to form a probable conjecture of the event. Something certainly may be gained if we obferve with attention. Thus I may remark, that I never found anticipations of one hour or even two to be of much
confequence in the fevers of Jamaica, particularly if
they happened at an early period; yet if they were
longer, or did not happen till after a long continuance of the difeafe, they often indicated an approaching crifis. On the contrary, where the paroxyfm anticipated twelve or fourteen hours at one time, there
was always fufpicion of danger, at whatever period
this might happen. It either indicated danger and
malignity, or a difpofition in the fever to change to a
continued form. Anticipating fevers were likewife
obferved to be more difpofed to terminate of themfelves, and likewife to terminate more fpeedily, than
thofe which fteadily preferved the fame hour of return. This feems to have been known to the ancients.—But farther, as anticipating types are generally a fign of increafing violence, though of a more
fpeedy termination; fo the poftponing of the paroxyfm, has ufually been allowed to indicate a difeafe,
whofe violence has begun to decline. Such is the
common obfervation; nor have I ever found it to be
otherwife; unlefs in fome cafes of weaknefs and im-

paired fenfibility, where the fatal paroxyfm did not come on till after the ufual hour of attack.

I am afraid I may appear to many as unneceffarily minute, on a fubject, which is not in general confi-dered as of much importance; yet ftill I cannot for-hear to mention fome circumftances of connection, between the hour of invafion and the future type of the fever, which appear to be not only curious, but ufeful; and which, fo far as I know, have not been taken notice of by any preceding writer. Galen, it is true, boafts the knowledge of foretelling, from the appearance of the firft paroxyfm, the nature of the fu-ture type of the difeafe; yet the hour of invafion is not included among the number of thofe figns, from which he has drawn his information. The rules, in-deed, which he has left us are not altogether without foundation; yet they are by no means to be depended on alone. They have, in fact, arifen from theories of bile and phlegm, rather than from accurate and careful obfervation. There is not, perhaps, any one crite-rion on this fubject that applies equally in every cli-mate. Thofe rules, which I thought I had difco-vered, are only local. They varied in the different parts of Jamaica, and in moft parts of America did not apply in any degree. Having adopted, on my arrival in the Weft Indies, the method of writing down and analyzing fuch cafes of fever as came un-der my care, the difference of hour, at which fevers of a different type ufually commenced, ftruck me as a matter of no fmall curiofity. The fingle tertian, for inftance, was conftantly remarked to begin in the forenoon, ufually between the hours of eight and eleven; whilft thofe forms of difeafe, that were evi-dently quotidian, or ftill more continued as con-ftantly began in the evening, generally from four to eight. This feemed to be a fixed diftinction; it is an important one; and fo far there was no ambiguity: but it was extremely difficult, perhaps impoffible, to

difcover figns at an early period, which were capable
of diftinguifhing the fever, which continued fimple
in its form throughout, or which became complicated
at a certain period of its courfe.   I have often made
a fortunate conjecture; but I believe it would occafion
embarraffinent, rather than afford information, were
I to attempt to defcribe thofe circumftances, which
fometimes determined my opinion.   They muft, in
fact, be learnt from actual obfervation.   I have juft
now mentioned, that it is extremely difficult to diftinguifh the fingle from the double tertian, by the appearances of the firft paroxyfm, when the morning
fever of this laft form begins the difeafe; fo neither is
it eafy to diftinguifh the double tertian from the quotidian, when the illnefs commences with the evening
paroxyfm of that complicated type.   This is a cafe,
indeed, as far as my experience goes, rarely happens;
yet where it does happen, the circumftances, with
which the evening fever is ufually attended, mark a
further diftinction.   The quotidian commences ufually by a diftinct and regular paroxyfm; the evening
fever of the double tertian, for the moft part, only by
a flight indifpofition.

This connexion which I have juft mentioned, between the hour of invafion and the type of the fever,
was obferved conftantly at Savanna la Mar in Jamaica;
but the fame rules did not by any means hold true in
the different parts of the continent of America.   The
moft ufual hour of the invafion of the fingle tertian
was twelve at noon in that country; though in fome
cafes the paroxyfin came on fo early as ten in the
morning, or fo late as two in the afternoon.   Of the
other forms I cannot fpeak with any certainty.

But befides the different hour of invafion of the
different types, I muft likewife take notice of fome
other circumftances, that feemed to be connected
with the various forms.   The duration of the paroxyfin for inftance, was ufually longer in the fingle
tertian than in the double tertian, or quotidian; and

of the double tertian, the paroxyfm of the morning
was ufually longer than that of the evening; and befides being longer, was generally of greater violence,
at leaft in the beginning of the difeafe, The cold fit
was likewife more remarkable in the fingle tertian,
than in the other forms of fever.—I do not fpeak of
the quartan, of which I know but little. Of the bilious vomiting, fo much infifted on by Galen as a
diftinguifhing mark of the fingle tertian, I am at a
lofs to fpeak pofitively. I can, indeed, affirm that I
conftantly obferved fuch evacuations to be more frequent in the different forms of the tertian, than in
thofe that appeared to be quotidian, or that approached
nearer to a continued type.

Such were the types of the fever of Jamaica, and
fuch were the principal circumftances that appeared
to be connected with them. The types of fevers it
may be obferved feem to be modified by climate; and
there are not perhaps two different countries, or even
diftricts of country, in which they are exactly alike.
The hour of invafion of the fame form of difeafe was
different in Jamaica, and on the continent of North
America; neither does it appear to have been exactly
the fame in Jamaica, and in the different iflands of
the Mediterranean. The prevalence of certain forms
in certain climates, and the various changes of the
form according to the changes of the feafon, is a matter of curiofity, and of confiderable importance in the
hiftory of the difeafe. In Jamaica for inftance, for
one fingle tertian, there were at leaft three double
ones; whilft in America, the fingle tertian bore the
proportion of ten to one, perhaps, to all the other
forms. In the higher latitudes of this country, the
fingle tertian was almoft the only form which was
feen, in the winter months and in fpring; yet in the
fouthern provinces, particularly in the heat of fummer and fometimes in autumn, the double tertian, and
even ftill more complicated types were by no means
uncommon.

I have thus endeavoured in the preceding pages to deſcribe the types of remitting fevers, as they occured to my own obſervation. I have likewiſe attempted to compare my own deſcriptions with thoſe of other authors; ſo that the reader may, in ſome degree, be able to attain a ſyſtematic view of this ſpecies of diſeaſe, as it appears in the different parts of the world. The labour I am aware will be reckoned ſuperfluous by many; and a diſcuſſion on types, will probably be conſidered as partaking too much of the ſchool of Galen, who is held in contempt by the writers of the preſent day. I by no means contend for the infallibility of Galen or the ancients; yet I cannot help believing, that though they have left us much falſe and ſuperfluous theory, they have likewiſe left us many valuable obſervations, on the nature and progreſs of febrile diſeaſes, which the moderns ſeem to have neglected. Though the theory of Galen on the preſent ſubject is probably ill founded, his obſervations are certainly exact; and from what I have myſelf ſeen, no aſſertions, which will convince me, that an intimate acqaintance with the types and periods of fevers, is not an eſſential knowledge to the practitioner. It is, indeed, the firſt ſtep; and it is a ſtep of ſuch importance, that our future progreſs will neither be ſatisfactory nor ſafe, unleſs it is properly underſtood. I may be allowed to ſpeak from my own experience. It fell to my lot to have the charge of men's lives at an early period of life. I had then no knowledge of the types and periods of febrile diſeaſes, and I muſt not conceal, that the method of treatment I purſued, and the returns of the diſeaſe ſo often interfered, that though actual harm was ſeldom done, yet many opportunities of doing good were certainly loſt. The patient, if he had the leaſt penetration, could not, ſometimes, avoid ſeeing, that, though I might be acquainted with the common routine of practice, I was ignorant of the nature and progreſs of the diſeaſe, from which I had undertaken to relieve him.

# CHAP. III.

THE critical days, which are intimately connected with the types and periods of fevers, come properly to be confidered in this place. The fubject is important, and though very fully difcuffed by medical writers, does not as yet appear to have been fatisfactorily explained by any one. If we attempt to trace the doctrine to its fource, we fhall find the firft mention of it in the writings of Hippocrates. The followers of this author's opinions are numerous; and the endeavours, they have employed in attempting to eftablifh his fyftem, have been ftrenuous and unremitting. But critical days have unfortunately afforded a field for controverfy in all ages; and there ftill are many who maintain, and others who as confinently deny the exiftence of any fuch power in affecting the termination of febrile difeafes. In the midft of this perplexity, a man who has had no experience of his own is totally at a lofs, which way to turn. There are great names on both fides of the queftion, but clear and decifive evidence on neither. A detail, therefore, of fuch obfervations as promife to remove many of thofe circumftances of embarraffment, cannot fail of being ufeful, and I hope acceptable to the public. I am aware, indeed, that an attempt to explain a myftery, which has hitherto eluded the refearches of the moft eminent phyficians, will not probably efcape cenfure:—In 'an obfcure man, perhaps, it may be deemed arrogant; neither is it altogether without hefitation that I undertake the difcuffion; though very unequivocal proofs of the truth of . the principles on which I proceed, arife from a view

of the cafes of fever which came under my care, during the time I lived in the Weft Indies. The principle on which the critical days in that climate depend,
appears from the facts found in the following pages
to be indifputably eftablifhed; the deviations are fatisfactorily accounted for; and the fundamental rules,
it is prefumed, are fuch as may be extended to every
climate on the globe. But, I fhall ftate in a few words
the leading circumftances, which occurred to me on
the fubject. If they afford not light enough to remove
all the difficulties, it is hoped they may at leaft point
out a road, by which thefe difficulties may in future
be removed.

It may not be improper to remark in the firft place,
that I had heard of the doctrine of critical days in fevers before my arrival in the Weft Indies; yet I may
likewife obferve, that it was a doctrine, which I had
only heard of by name. I had no knowledge of it,
and I foon found that the idea was treated with ridicule by practitioners, who very generally fuppofed'
the courfe of the fevers of Jamaica to be cut fhort by
bark, or other powerful means. Influenced, perhaps,
by the authority of older men, I found myfelf difpofed
to acquiefce in the common opinion, that this doctrine
was only one of the fanciful theories of the fchools;
yet it was not long before I acquired a different view
of the fubject. I foon obferved that fevers fometimes
ceafed before a fingle grain of bark was given;
fometimes after a few dofes, and fometimes not
till after feveral ounces. The obfervation of this
fact did not fail to undeceive me. Under the circumftances I mention, it would have been the height
of obftinacy to have perfifted in believing, that the
cure of the fever was in reality owing to the power
of this celebrated remedy. But though it was foon
evident, that the termination of the difeafe depended
on fome other thing than that which was generally
fuppofed ; yet a confiderable time paft over, before I
was able to determine what this fomething actually

was, or before I was able to afcertain the laws which
it obeyed. I foon difcovered, indeed, that fevers had a
general tendency to terminate on particular days; but
it was not till the year 1776, that I difcovered the
proportion thofe days bore to each other, or the fources
of the many deviations, which occurred. The fubject
feemed to be important;—and I felt an eagernefs to
determine a queltion, which hitherto had been fruit-
lefsly purfued. With this view I wrote down with
care and attention every cafe of fever which I met
with in the year 1776 and 1777; and in looking over
the memoranda fometime after, found, that the critical
days bore to each other the following proportion. viz.
of fixty cafes, which terminated favourably, ten terminated on the third, ten on the fifth, twenty on the
feventh, ten on the ninth, five on the eleventh, three
on the thirteenth, and two on the feventeenth. Of
nine which terminated fatally, one terminated on the
fixth, one on the feventh, fix on the eighth, and one
on the tenth. Thefe facts are precife and determinate; but I muft not forget to mention, that if we are
guided wholly by obvious appearances, there fometimes occur circumftances, which occafion embarraffment. Thus in the prefent inftances, I not only reckoned the time by the periods, or revolutions of the
difeafe, but I likewife fimplified the complicated types;
that is, I reckoned every revolution of the fingle tertian as forty-eight hours, though it was often com-
pleated in lefs; whilft I confidered the correfponding
paroxyfms of the double tertian, only as the fame difeafe. It may alfo be farther remarked, that the difeafe,
which was the fubject of this inveftigation, was of a
regular, remitting form. Paroxyfms and remiffions
were always difcernible, and figns of crifis were generally diftinct.

The ftate of the critical days, as reprefented above
is literally exact, where the complicated types were
fimplified, and where the time was reckoned by the periods of the difeafe; but I muft likewife obferve, that

unlefs this method of calculation was adopted, there cc-
cured numerous inftances, which feemed to deviate
from the general rule. In the firft place, if the type of
the fever was fingle tertian, which neither anticipated
nor poftponed,—and with paroxyfms which did not
exceed twelve hours in duration, the crifis was uniformly on an odd day: yet if the type anticipated, and
the fum of the anticipations, in the courfe of the difeafe, was equal to twenty-four hours, the crifis was
then necessarily removed to an even day, if the time was
reckoned by the natural day; though ftill on an odd
day, if reckoned in the manner which has been mentioned above. In like manner, if the type poftponed,
while the duration of the paroxyfm exceeded or
amounted to 24 hours, the crifis was neceflarily
protracted to an even day. But this was a cafe, which
feldom happened. In fevers likewife of the double
tertian type, the type which prevailed principally at
Savanna la Mar, there occurred much feeming irregularity. This form of fever, as was faid before,
feemed to confift of two difeafes, which ran a feparate
and independent courfe. Thus, if the fever which
began on the odd day was critical; that is, if the paroxyfm of the odd day terminated the difeafe, the crifis was neceflarily on an odd day; but if that fever,
the firft attack of which was on the even day, confifted
of an equal number of paroxyfms with the other, or
continued after that had ceafed, the crifis was then on
an even day, reckoning from the beginning of the illnefs, though ftill on an odd day, dating from the
commencement of the fecond fever. It was the obfervation of this fact which firft gave me the idea of
fimplifying complicated types, and of calculating the
critical days by the periods of the difeafe. The idea
may perhaps be reckoned fanciful; but experience
has afforded me fufficient proofs, and it will ftill afford the fame to thofe who take the trouble to look
for them, that the various types of complicated fevers

actually run a feparate and independent courfe; a fact
when eftablifhed, which removes all doubt and ambiguity from the apparently varying laws of critical
days in the compound forms of febrile difeafes. With
regard to the quotidian it remains to be remarked,
that the crifis was generally on an odd day. It was
likewife generally on an odd day in thofe that were
ftill more continued and acute;—a fact which feems
to have been well known to Avicenna. But though the
rules I have mentioned are clear and uniform, I muft
ftill own, that I have fometimes met with fevers of a
very continued kind, which terminated late on the
fixth, or rather very early on the feventh. The difeafe
was then of more than ufual violence on the fixth:—
how far this might be owing to anticipations of the
paroxyfm of the feventh, accumulated upon that of the
fixth, is difficult to determine with certainty.

The anticipation, the poftponing, and the complication of type are the principal circumftances, which
ufually difturb the regular critical periods in fevers of
fhort duration; yet in thofe of longer continuance,
there is ftill another caufe, which deferves to be particularly attended to. In the fevers of Jamaica, efpecially in thofe which approached to a continued form,
fome very apparent change in the nature of the fymptoms, or in the mode of action of the febrile caufe,
was generally obferved on the feventh, or before it.
In confequence of this change, the order of the critical days was fometimes difturbed, and appearances
were often produced, which feemed to contradict the
rules, which we have attempted to eftablifh. It was
a common remark, that after the feventh there was
lefs apparent regularity in the movements of nature,
This, as we fhall afterwards attempt to prove, was
the confequence of a feptenary revolution, which
accidentally difturbed the regular order of the ordinary days of crifis. It is a fact of which the ancients
were not ignorant; and of which I fhall have occa-

D

fion to make frequent ufe: viz. that a relapfe has a tendency to run over a courfe of duration equal to the original fever. This is confirmed by the authority of Hippocrates; but I may alfo add, that not only thofe recurrences of fever, which are more properly ftyled relapfes; but further, that in thofe inftances, where the difeafe undergoes any remarkable change in the nature of its fymptoms, the diforder is generally difpofed to continue for the fame length of time in this new form, as it had done in the former. Thus a remarkable change of fymptoms on the fifth was followed by a crifis on the ninth; fometimes, perhaps, only by another change of fymptoms on the ninth, the final crifis not happening till after another period of five days. In like manner, a change of fymptoms on the feventh was often followed by a crifis on the thirteenth; or only, perhaps, by another change on the thirteenth, the difeafe completing another revolution of feven days before a final termination. That fuch changes actually do take place at certain periods, not only thofe cafes of fever, which have come under my own care, but thofe related by Hippocrates, in the books of Epidemics, give fufficient room to believe. Thus in every one of thofe inftances, where the hiftory is fo circumftantially detailed as to leave it in our power to trace the difeafe in its progrefs, it will conftantly be found, if the day of crifis deviates from the general rule, that a change of fymptoms, often an evident renewal of fever, had actually taken place at fome period of the courfe. In this manner, if the change of fymptoms of which I fpeak happened on an odd day, the odd days continued to be critical, as if no change had been; on the contrary, if the paroxyfm of the odd day completed its courfe, the remiffion which followed was often more perfect than ufual :——a diftinct period was marked in the hiftory of the difeafe,——or in other words, there was an obfcure or imperfect crifis. But on the day following, which was an even day, a fever with a different train of fymptoms made its appearance, and ran over a courfe, for the moft part, equal in duration to the former. If this change, or renewal of the difeafe happened on the fixth, a change or crifis was not expected till the tenth, if on the eighth, not till the fourteenth. I have faid juft now, that relapfes were generally difpofed to run over a courfe of the fame duration as the original difeafe ; yet I muft likewife remark, that they were fometimes alfo of fhorter continuance. Thus I have frequently obferved a change of the nature of the fymptoms on the feventh, and a final crifis on the eleventh ; the renewal of the difeafe, inftead of feven, being only of five days continuance.

The above circumftances are capable of explaining the ordinary deviations from the regular critical periods in the fevers of the Weft-Indies; but I cannot affirm with the fame certainty, that a fimilar explanation will be conftantly admitted in the long fevers of this country. I have however reafon to believe, that changes at the feptenary periods frequently take place here, and fometimes apparently difturb the critical periods of the difeafe. Thofe cafes which I have been able to trace with accuracy give ftrong proofs of it,——I fhall relate two or three of them to ferve as an illuftration. The firft, is that of a young man, who had been ill of a fever more than three weeks before I was called to him. Two days before I faw him ; and after an evident abatement of the fymptoms, there happened a fudden and unexpected relapfe, or renewal of the difeafe. Informed of this circumftance, I dated from the new attack, and calculated the critical days in the manner which has been fhewn above. Minute attention difcovered the type, though it was only an obfcure one.——It was Semitertian, or there was an exacerbation every evening, with a more evident paroxyfm

on the alternate days. A crifis happened at the period
I had forefeen, but it was not final. A fever returned
again in the evening, different however in type, as
well as in fymptoms, from the preceding. It had
diftinct quotidian exacerbations and an imperfect crifis
happened on the feventh. But in twelve or fourteen
hours, a coldnefs and fhivering marked a renewal of
the old, or perhaps the invafion of a new difeafe. The
fymptoms were not only different in their nature from
the fymptoms of the former; but they were likewife
more violent in degree. The difeafe continued in
this form for feven days, and the crifis, which at
laft was only imperfect, was foon fucceeded by another
renewal of fever, the beginning of which was marked
by a fimilar degree of coldnefs and fhivering. The
fymptoms of this were likewife different from the preceding, but its form was the fame, and it ran over a
courfe of equal duration. The feptenary revolutions
were very plain in this cafe. I fhall relate another in
which they were not fo clearly marked, though they
certainly did ftill take place. It is a cafe of fever with
nervous fymptoms. On the feventh a fediment appeared in the urine, fome drops of blood fell from the
nofe; and the abatement of fever was very evident;
yet it did not laft long. The difeafe recurred again
on the eighth, and continued to increafe in violence
till the fourteenth. A fediment then appeared in the
urine, fome drops of blood fell from the nofe as before, there were two or three evacuations by ftool,
which had been unufual in the preceding courfe of the
difeafe; and from the whole appearances I could not
help entertaining fome faint hopes of crifis. There
was indeed an evident alleviation of the fufferings;
but it lafted but for a fhort time. Next day every
fymptom was aggravated, and the powers of life feemed to fuffer a gradual diminution till the twentieth,
when the patient died. I do not recollect any inftance of fever, where the revolutions were more ob-

fcure than in the prefent cafe; yet they were ftill capable of being traced. The next example I fhall mention is much clearer. It is an inftance of a bad fever, of no difcernible type in the beginning, in a man who was confiderably advanced in years. On the evening of the feventh there was fome obfcure tendency to crifis. The patient was not only eafier in his own feelings; but the eye and countenance, which had been confufed and clouded, brightened up, and a fmall fediment appeared in the urine. Yet thefe favourable circumftances were only of fhort duration. In the courfe of the day following, all the fymptoms recurred, and the difeafe acquired force till the evening of the thirteenth. The pulfe then began to rife, and continued rifing till the morning of the fourteenth, when a profufe fweat was followed by a very diftinct crifis. But ftill this crifis was not final. The malignity of the difeafe, however, departed, and the complaint that remained, affuming a remitting form, totally difappeared after another period of feven days. I fhall only beg leave to relate another inftance of fever, which occured to me lately, and which affords a very curious proof of feptenary revolutions in febrile difeafes of long continuance. A young man had been ill of a fever about a fortnight before I was called to him. At the time I firft faw him, the fymptoms were very violent; but having abated confiderably in the courfe of a day or two, I began to entertain hopes of a fpeedy recovery. The complaint was almoft entirely gone, when a new train of fymptoms unexpectedly making its appearance, raged with violence for a day or two, and then declined gradually as the other had done. I again looked for figns of crifis, when another acceffion on the feventh from the former attack, brought matters into ftill greater danger. Thefe fymptoms, though of a different nature from the former, were violent in the beginning; but they foon began to abate, and had almoft difappeared, when the attack was once more renewed on the following feventh. In this manner the difeafe went through nine feptenary revolutions; and it is fomewhat remarkable, that the fymptoms, which marked the new acceffion, were always different from thofe of the acceffion immediately preceding. In one, the diftinguifhing fymptoms were a morofe and ftern fullennefs, in another, delirium, tremors and fubfultus tendinum,—and in the third, copious liver-coloured ftools. Thefe were three times feverally repeated. It deferves, however, to be re-' marked, that the period of the acceffions was fhortened before the termination of the difeafe. After it had continued nine weeks in the manner I have deferibed above, there were two acceffions of five days each; after which all traces of fever difappeared.

It is fufficiently plain from the facts which I have mentioned in the preceding pages, that the more ufual irregularities in the order of the critical days, proceed generally from overlooking the type in periodical fevers, or from neglecting to attend to feptenary, and other revolutions, in fuch as approach more nearly to a continued form. Thefe are the general caufes of apparent irregularity; yet befides thefe, there are ftill fome others, which muft not be paffed over without notice, as they occafionally have the effect of producing apparent deviations. Thus it often happens, that a difeafe, which appears to be continued in the beginning, changes to remitting after a certain duration. The change is ufually on an odd day, and on the day following the firft paroxyfin of the remitting form makes its appearance, the termination of which may be expected on an even day, if we date from the beginning of the illnefs, though ftill on an odd day, if we date, (as perhaps we ought to do) from the time this change in the circumftances of the difeafe took place. To this we may add, that thofe complicating fevers, which, happening at various dif-

tances of time, fometimes terminate fooner, fometimes continue longer.than the original complaint,
frequently difturb in appearance the general regularity of the critical periods of nature. It happens,
perhaps, from a fimilar caufe, that a paroxyfm of an
unufual kind fometimes terminates the difeafe, and
apparently difturbs the regular periods of crifis This
has occurred to me feveral times in practice; and it
happened twice in my own perfon. The ordinary
paroxyfm declined after the ufual duration; a new
one fucceeded of uncommon violence, and very different in its nature from the former. Its courfe was
of long continuance, and it finally terminated the
difeafe.

The above facts enable us to explain fatisfactorily
every circumftance, which relates to critical days in
fevers, where the crifis is clear and decided; yet I
muft ftill own, that as I have fometimes met with
fevers where marks of crifis were fcarcely perceptible
fo it would be rafhnefs, in fuch cafes, to fpeak pofitively of the order of the critical days. The patient
might, in fome meafure, be faid to wade through the
difeafe; the changes from day to day being fo very
fmall, that it required more difcernment than I can
boaft of to mark them with precifion. .

The obfervations I have related, and the rules I
have attempted to eftablifh, for the better explanation of the doctrine of critical days in fevers, were
formed at a time when I had no knowledge of the
opinions of preceding authors. They may therefore
better claim exemption from bias in favour of one
fet of writers, or prejudice againft another. They
are indeed no more than an analyfis of facts, which
were collected with every poffible care, which are
fufficiently circumftantial, and which fpeak beft for
themfelves. They contain, (if I do not view them
with a partial eye,) fuch information, as may lead to
a fatisfactory explanation of this myfterious and long

difputed doctrine.—I muft only beg leave to add, that though I have everywhere mentioned the pre-eminence of particular days in terminating fevers, yet it muft not be underftood, that this power depends on a particular quality of the days, merely as fuch. It depends more evidently on a certain number of revolutions of the difeafe, in confequence of which, the fever from fomething we do not in the leaft underftand, feems difpofed to terminate finally, or to fuffer a change in its mode of action. This therefore brings us to the conclufion, that the critical periods are improper calculated by the natural day. The doctrine, in fhort, can only be rendered confiftent by attending to the periods of the difeafe, by fimplifying complicated types, and by marking thofe feptenary or other revolutions, which happening at different diftances of time, occafion an appearance of irregularity which does not exift in reality.

Having related the refult of my own obfervations on critical days in fevers, I fhall now endeavour to bring under one point of view, the fubftance of what has been written on the fubject, by fome of the moft celebrated of the ancient, as well as modern phyficians. That certain days, or that portions of time comprehended in a certain number of days, had obvioufly a power of producing changes on the human frame, appears to be an obfervation of high antiquity; but as a medical doctrine, we are unable to trace it farther than the days of Hippocrates. Hippocrates has treated very fully of the critical periods of fevers, in various parts of his works; and upon the whole, has amaffed a confiderable body of information; though with lefs precifion, perhaps, than has been generally imagined. The cafes of the Epidemics, which we naturally confider as the materials from which he formed his general doctrine, have fome obvious and great defects. The date is feldom clearly afcertained, and the mode of calculating the time, does not feem to be fixed. If

a fever, for inftance, begins in the evening, or in the
courfe of the night, the day following is generally
reckoned the firft day of the difeafe, by this author.—
But this is not all.—Some of the cafes are plainly related from memory; and others are only parts of
cafes, related by different perfons. This want of
accuracy, where it is fcarcely poffible to be too circumftantial, neceffarily breeds confufion, and produces an appearance of irregularity, which does not
actually exift. Hence we find inconfiftency in the
general doctrine, as delivered in different parts of the
works, which have been afcribed to Hippocrates; at the
fame time, that there is a want of that circumftantial
detail in the particular parts, from which only we can
be enabled to form an opinion. I have read over
with much attention the cafes of fevers, recorded in
the Epidemics; but I frequently found myfelf unable
to trace the difeafe in its progrefs. Though evidently
fubject to periodical movements, it was not always
in my power to lay hold of the type; yet wherever
it was poffible to attain this exactnefs, I have the
fatisfaction to add, that I conftantly found the movements of nature to be uniform. They were the fame
in the iflands of the Archipelago, as in the ifland of
Jamaica.—If they appeared in fome inftances to be
different, it was perhaps principally owing to this,
that the Greek phyfician had left fome part of the difeafe undefcribed.

· From what I have juft now faid, we can have no
hefitation in concluding, that the opinion of Hippocrates, on the fubject of critical days, is neither precife in any one part, nor confiftent in the whole. The
doctrine, however, in its beft digefted form, is the
following: viz. That odd days have a remarkable
power in terminating fevers; but more particularly,
that the great critical revolutions happen at quaternary periods. Thus the moft eminent critical days,
are the fourth, the feventh, the eleventh, the four-

teenth, the feventeenth and the twentieth. This is the general form of this Hippocratic doctrine; yet in this form, it bears contradiction to obfervations that are found in various parts of that ancient author's works. The fifth and ninth are excluded by this arrangement, from the number of the critical days; though there are numerous examples of their great power, in terminating febrile difeafes.

The doctrine of critical days, which appeared firft in a regular form, in the writings of Hippocrates, found numerous and refpectable advocates among the ancient phyficians. Diocles of Caryftus, Philotimus, Heraclides of Tarentum, &c. all bore teftimony to the general truth of the obfervation; but their writings being unfortunately loft, we are now ignorant of the particular facts and arguments, by which they attempted to fupport their opinions. Indeed, from the time of the Perfian invafion of Greece, till the Roman arms penetrated into Afia, a period of near four hundred years, we know of no oppofition to to this fundamental doctrine of the Coan Sage : But in the time of Pompey the Great, an author arofe, who endeavoured to eftablifh his own fame on the ruins of this favourite fyftem of his predeceffors. Afclepiades, who was a man of a bold and daring genius, not only rejected this apparently well founded doctrine of the ancients, but treated the idea of it with ridicule. His arguments are ingenious and acute; but they fall fhort of the truth. The paroxyfms or exacerbations, as he juftly obferves, fometimes change to the even days, and confequently the crifis: yet this, if properly underftood, does not deftroy the generality of the rules;——if the method of calculating the time, by the periods and revolutions of the difeafe, be adopted, the difficulty is perfectly removed. But though this fact in reality, was not unknown to Afclepiades; yet it does not appear, that he underftood the application of it. I may add, that he has precipiately rejected the doctrine, from the very circumstance which establishes its reality.

We do not meet with any thing very material, on the present subject, between the time of Asclepiades, and the days of Galen. There appears, indeed, to have been many, who adopting the opinion, and copying the arguments of the eloquent Bethynian, denied altogether the existence of critical periods in fevers; whilst others, recurring to the doctrine of Hippocrates, maintained their reality with no less obstinacy. But we are now in a great measure ignorant, if those writers attempted to support their opinions by any new facts, or new arguments. Among other misfortunes, we must regret particularly, that the treatise of Aretæus on fevers is lost. From what we know of this author's industry we might have reasonably expected original information on the subject in question.

Galen, whose fertile and exuberant genius left no path in physic unexplored, has written fully on this celebrated doctrine. He has professedly adopted the opinion of Hippocrates, and laboured much to explain and confirm it; but unfortunately, he has oftener overwhelmed the subject with diffuse and tedious reasonings, than illustrated it by proofs from experience and actual observation. Upon the whole, however, amidst much superfluous and unmeaning matter, we find not only useful information, but a more systematic arrangement of facts, than is any where to be met with. He has attempted to fix with more precision the date of invasion; he has estimated with more accuracy the critical power of the different days; and further, has hinted obscurely, that the time will be calculated most conveniently by the paroxysms or revolutions of the disease. In short, this author, no less than Asclepiades, was sufficiently acquainted with the principal truths, which give consistency to this doctrine; but it is evident, that he did not underſtand the full extent of their application. He was conſtantly biaſſed by the theory of a quaternary period; as without this predilection, it is not eaſy to conceive, how he ſhould have conſidered the fourteenth, as critical of tertians, where the paroxyſms happen on the odd days, and where the termination, as he acknowledges, conſtantly follows the ſolution of a paroxyſm. The latitude likewiſe which he aſſumes, in explaining the apparent irregularities, is much too great. If we are permitted to reckon either the beginning or the termination of a paroxyſm, as the critical period, according as it ſhall beſt ſuit our theory, it is eaſy to elude the moſt poſitive teſtimonies of experience. Yet, notwithſtanding theſe defects, the different tracts of Galen on this ſubject, deſerve to be carefully read. The facts they contain, though ſometimes miſapplied, are often important; and though we are not always ſatisfied with the reaſonings of the author, we are aſtoniſhed at the amazing maſs of learning and knowledge found in his works.

There is little new information, on the ſubject of critical days, to be met with in the writings of thoſe Greek phyſicians, who were poſterior to the time of Galen. Ætius Amidenus, indeed, brings into narrower compaſs the ſubſtance of the doctrines of his predeceſſors. He mentions likewiſe, the moſt material of thoſe circumſtances, which influence the deviations from the regular criſis ; but it is evident, that he has not ſufficiently underſtood their application. Alexander Frallianus, who was an excellent practitioner, and a man of long experience, paſſes over this ſubject without particular notice ; and though Paulus of Ægina has detailed the opinions of Galen in a more compreſſed form, than they are found in the original author ; yet he has not added any new obſervations of his own. From the manner, indeed, in which he ſpeaks, of the peculiar virtue of the feventh and fourteenth, we fhould be apt to believe, that he is not altogether free from prepoffeffion in favour of the Pythagorean numbers.

It was reafonable to have expected information, on the fubject of critical days, from the writings of the Arabian phyficians. The Arabians inhabit a country, were the periodical movements of nature are perhaps more clearly marked, than in our northern latitudes. Some diftricts of their country likewife were famous for the fciences at an early period, though it does not indeed appear that much of this knowledge defcended in a direct channel to the Arabians of the prefent times. The Arabian phyficians, in many inftances, enriched medical practice with new forms of remedies; but they have for the moft part only adopted the theoretical doctrines of the Greeks, particularly of Galen. Avicenna, the moft famous among their phyficians, and undoubtedly a great man, has Galen conftantly in his eye: in fhort, he has done little more on the fubject of critical days, at leaft, than merely tranflate the opinions and arguments of the celebrated Greek. He attempts, indeed, to be more explicit in afcertaining the date of invafion; but he does not in fact, go much beyond his predeceffors;—hinting only obfcurely, that the critical days ought to be calculated from the proper formation of the type, or the diftinct invafion of the fever. He has added, however, that the odd days, are properly the critical days of the fingle tertian, and that the eleventh of courfe, obtains rank of the fourteenth in this difeafe.

There are many authors, who have written on this fubject, fince the arrival of fcience in Europe; but there are few that I have met with, who have thrown light on it from their own obfervations. The moft of them have borrowed the opinion from Hippocrates; and accordingly have attempted to eftablifh the truth of it, on the facts which are found in

E

the writings of that author; facts, which, on enquiry, will scarcely be found to be accurate enough to be made the basis of a general doctrine. It would be time ill spent, to enter into a detail of the arguments of this numerous lift of writers; who, in reality, have oftener attempted to support their opinions by the authority of Galen and the ancients, than by the facts which might have been found in their own experience. From writers, however, of this description, it would be unjust, not to separate Hoffman, an author, who has related with candour the result of his own observations, in a practice of forty years and upwards. The facts which Hoffman mentions, throw considerable light on the subject; yet still they do not remove all the difficulty. They neither enable us to form an estimate of the power of the different critical days; neither do they at all assist us in comprehending the caufe of the deviations. There are probably other modern authors besides Hoffman, who have treated of the power of critical days in fevers; but, except Dr. Cullen, I have not met with any one, who has left any observations which deferve much notice. This celebrated phyfician is a warm advocate of the ancient doctrine of critical days. He subscribes profeffedly to the arrangements of Hippocrates; though he adds likewife the result of his own observation, in the various kinds of fevers of this country.

The moft eminent of the ancient and the moft fyftematic of the modern phyficians, all agree in aferibing to certain days a particular power in terminating; yet they do not fo perfectly coincide in the arrangement they have given of thofe days, or in the caufes they have affigned for the particular pre-eminence. The inconfiftency of Hippocrates has, perhaps, been in fome meafure the fource of this diverfity of opinion. In one place, this author has ranked the twentieth as the proper critical day in fevers; in fome others, this power is attributed to the twenty-

firft. That the twenty-firft is properly the day of
crifis, was the opinion of Archegenes and Diocles;
that it fhould be fo, is not inconfiftent with the general principle of the Hippocratic doctrine; viz. the
movements of a quaternary period. So far is clear;
but as it was obferved by Hippocrates, as well as by
other authors, that the twentieth was ftill more frequently a day of crifis than the twenty-firft, a προσθεσις,
on the fourteenth, was introduced to account for this
apparent deviation from the general rule. This idea
of προσθεσις, or accumulation of one period on another, which is mentioned in the writings of Hippocrates, originated perhaps in the doctrine of Pythagoras. It is adopted by Galen, and it appears in
reality to be occafionally true; yet it can never be
confidered as an eftablifhed principle in the movements of febrile difeafes. By means of fuch accumulation, however, Galen has attempted to eftablifh
the pre-eminence of the twentieth, which he confiders as the real critical day of Hippocrates. That
the twentieth—(not the twenty-firft) is actually the
critical day of Hippocrates, is likewife decidedly the
opinion of Dr. Cullen, who, going a ftep farther
than his predeceffors, endeavours to fupport his affertion by fome arguments, which are entirely new.
This ingenious author hazards the bold conjecture,
that the appearance of the twenty-firft, in the writings
of Hippocrates, has arifen wholly from accidental
error in the original manufcript: but with all due
deference to fuch refpectable authority, I muft beg
leave to fuggeft, that the twenty-firft occurs too frequently in thofe writings, which have been afcribed
to the Coan Sage, to give countenance to the opinion, that it owes its place, as a critical day, to care-
lefs error. The other argument is more ingenious;
but perhaps not better founded. This writer has
ventured to maintain, that the type of febrile difeafes
changes to quartan after the eleventh; but I can fee

E 2

no good reafon for the fuppofition. Medical writers have repeatedly noticed inftances of crifis, on the thirteenth, and fifteenth; even my own experience, narrow as it has been, furnifhes me with fufficient evidence, that crifis actually do happen at the above-mentioned periods.

As thofe days, which have been chiefly confidered as critical, are now fuppofed to be fufficiently known, it will not be fuperfluous in the next place, to take a fhort view of the caufes, on which the particular pre-eminence has been thought immediately to depend. The quaternary period, which in reality is a period of four, and a period of three days fucceeding each other alternately, is the general principle affumed by ancient phyficians, to explain this arrangement. But if we continue to purfue the undifturbed movements of a quaternary period, we fhall bring the eighteenth and twenty-firft into the order of critical days, rather than the feventeenth, and twentieth. The contrary is in fact the cafe. To obviate therefore this difficulty, or to reconcile obfervation with theory, a προσθεσις has been fuppofed to take place on the fourteenth. That a προσθεσις, or as it may be tranflated, the accumulation of the beginning of one period on the extremity of another, frequently takes place, cannot be denied; but its appearance is not determined by a fixed law. It is obferved on the feventh, on the fourteenth; in fhort, on any day whatever. The quaternary period, with προσθεσις on the fourteenth, is the only principle employed by the ancients for explaining the ufual arrangement of the critical days; yet I muft obferve, that it is capable of doing this, only in a very imperfect manner; it totally excludes fome days of very confiderable power. Dr. Cullen, fenfible, perhaps, of this defect, fuggefted that there was a change from the tertian to the quartan type on the eleventh. This change, it muft be confeffed, explains with perfect plaufibility

the pre-eminence of the fourteenth, feventeenth and
twentieth ; but there is the ftrongeft reafon to believe,
that it does not in fact take place. I mentioned before, that inftances are recorded by medical writers
of crifis, which have happened on the thirteenth, fir-
teenth, and the other days, which are not included in
the quartan period ; and I can add from my own experience, that where the difeafe was of fuch a kind,
that a type could be clearly traced ; no fuch change,
as this author has fuggefted, was ever feen.

Having ventured to declare, that the caufes,
which have been hitherto affigned for the pre-eminence of certain critical days in fevers,.are extremely
defective; the facts, which I have mentioned before,.
it is prefumed, may enable us, if they are properly
underftood, to give a more fatisfactory explanation
of this fingular phenomenon. There are few people
of experience and obfervation, who do not know that
the tertian is the moft prevailing type in febrile difeafes. This, at firft fight, gives a general pre-eminence
to the odd days; but though the tertian period prevails very generally in fevers, yet it muft alfo be remembered, that thefe revolutions are fometimes completed in a fhorter fpace of time than the regular period ; whilft the types are frequently found to be
doubled, or even more varioufly combined. In confequence of thefe accidents, apparent irregularities
are often produced in the order of the critical days;
though they may be all fatisfactorily accounted for,
by calculating the time by the periods of the difeafe,
or by fimplifying thofe types which are more evidently complicated. By attending to the circumftances I have mentioned, all the difficulties may be
eafily removed in periodical fevers; but as numerous
inftances of fevers occur, where no type can be clearly
traced ; fo it is neceffary in fuch cafes to feek for
fome other principle, which may be capable of explaining apparent irregularities. There very feldom.

E 3

perhaps happens an inftance of fever of long continuance, where the fymptoms do not undergo fome
change in the courfe of the difeafe. Thofe changes
or revolutions are generally at confiderable intervals,
frequently at an interval of feven days. The circumftances by which thofe changes are indicated, are not
by any means obfcure; and, perhaps, there would not be
great error, if we confidered them as the commencement of a new complaint; at leaft by confidering them
as fuch, the general principle of the critical days is preferved confiftent and uniform throughout. I fhall mention fuch explanations as have occurred moft frequently
in my own practice. It often happened, that the
fymptoms of the difeafe underwent a material change
on the fifth. It terminated on the ninth, or perhaps
only put on a new appearance on the ninth, its final
termination not happening till after another period
of five days. In the fame manner, a change of fymptoms on the feventh, was followed by a crifis on the
thirteenth; or if the change of fymptoms was not
obferved till the ninth, the crifis probably did not
make its appearance till the feventeenth. Such
change of fymptoms on the odd days, (where we may
fay with propriety enough, that one difeafe was accumulated upon another), there being feldom any previous marks of crifis, was by no means uncommon;
yet it happened ftill oftener, that the paroxyfm of the
odd day declined; the original difeafe terminated imperfectly, whilft a new one began the day following,
which was an even day. By fuch accidents the order of the days of crifis was changed: And from
the laft mentioned caufe the fourteenth, as a fecond feventh, becomes remarkable among the critical
periods of fevers. This idea of a fecond feventh occurred to me many years ago, and long before I was
acquainted with the opinions of Hippocrates or of
Galen. It now receives information from the teftimony of thefe careful obfervers. There are many,

I make no doubt, who will be difpofed to treat it with ridicule; but I fhall combat their opinion with no other argument than a requeft, that they write down carefully the hiftory of a tedious fever, and afterwards review its courfe without prejudice or partiality.

I now only beg leave to add, that the facts which I have mentioned in the preceding pages are circumftantial, and give room to conclude, that by fimplifying complicated types, by calculating the time by the revolutions of the difeafe, or by beginning to date a fecond time from thofe great and remarkable changes, which happen at more diftant periods, a doctrine is formed, perfectly uniform and confiftent with itfelf. It is confirmed by every obfervation which I have been hitherto able to make. It is no more indeed, than an analyfis of thofe feveral cafes, which have occurred in my own practice; which in periodical fevers at leaft, has been tolerably extenfive.

But though the prevalence of a tertian type, explains fatisfactorily the general critical power of the odd days; and thofe other circumftances, which I have likewife taken notice of, account no lefs clearly for all the deviations, which are obferved to take place; yet if we attempt to feek for a caufe of this type, or of thofe changes, which happen at more diftant, particularly at the feptenary periods, our progrefs is foon ftopt. Galen, who feldom hefitates in explaining the phenomena of nature, acknowledges here that he was unwillingly drawn to a difcuffion of the fubject. The queftion undoubtedly is a difficult one; and, it is to be feared, muft remain for ever unknown. In the Eaft, where the powers of the human mind were not only earlier developed; but where men, from climate and modes of life, were led more early to obferve the motions of nature, ftated and periodical movements were foon difcovered in the economy of the fublunary fyftem. Egypt,

there is reafon to believe, is one of the countries
where thefe revolutions were firft taken notice of; at
leaft it was on the banks of the Nile, that the Greek
philofophers firft gathered the feeds of natural fcience.
Among the knowledge or opinions, which thefe fages
carried back to their native country, we may reckon
the doctrine of the power of numbers; which though
disfigured perhaps by the metaphyfical genius of the
philofopher of Samos, has obfervation in fome degree
for its bafis. It does not concern us at prefent to
enter into a particular difcuffion of this opinion; but
as far as relates to the fubject in queftion, we cannot
refufe acknowledging, that the frame of man is liable
to regular changes, at particular periods, comprehended in a certain number of days and hours. But
though this general truth is indifputable, yet there is
no argument which leads us to fuppofe, that thofe
changes are, in any degree, influenced by an harmo-
nic proportion in the fimple number of the days. Ill
founded however as this doctrine obvioufly is, it
was in high fafhion with the Greeks, in the time of
Hippocrates; and feems evidently to have had fome
influence on the opinions of this author. Without
fuch a prepoffeffion, indeed, it is not eafy to conceive,
how he could have fabricated the fyftem which he has
given to the world; as it by no means refults from
the facts which are found in his writings. Galen in
this, as in moft fubjects, follows the footfteps of
Hippocrates. He difclaims, I muft confefs, the
*power of numbers*, fimply as numbers having any
effect upon the moft ufual days of crifis; but he
maintains the influence of a quaternary period, which
appears very plainly to be a remnant of the doctrine
of Pythagoras. However, after exhaufting himfelf,
and fatiguing his readers with a detail of ufelefs conjectures, he at laft ventures to conclude, that the bufinefs of crifis is to be referred ultimately to the
courfe and different afpects of the moon. The opi-

nion, like many others recorded by the Greek phyficians, draw its origin from Egypt. It is not, perhaps, altogether without appearance of plaufibility;
yet I muft add, that if the moon has in reality any
influence in this bufinefs, the laws which regulate
its effects are obfcure;—indeed, not in the leaft underftood. The conjecture however, fanciful as it
appears to be, met with the general affent of medical
writers, till about the middle of the fixteenth century, when Fracaftorious, a man of ingenuity and
elegant genius, attempted to fubftitute another in its
place; though unfortunately, not a more probable
one than that of his predeceffors. This author, after
a difplay of much learning and general knowledge,
at laft ventures to conclude, that the power of the
different days of crifis, depends on peculiarities in
the laws of motion of the different humours, which
give rife to the different fpecies of the difeafe : but
with regard to this hypothefis, it is only neceffary to
remark, that while the very exiftence of the humours
is doubted with reafon, there can be no certainty in
determining the laws of their motions. But though
the opinion of Galen, and this of Fracaftorius, are
only vague and very queftionable conjectures; yet
they are the only ones, fo far as I know, which have
been offered to the public. The fubject is too intricate, perhaps, ever to be explained. For though
we clearly perceive that fevers are ufually of a ftated
duration; yet we are unable to perceive, whether
this duration depends on fomething inexplicable in
the peculiar nature of the caufe, which ceafes to act,
or changes its mode of action at a certain period; or
to fome imperceptible revolution in the human frame,
which deftroys in a given fpace of time, that particular aptitude between the ftate of the body and the
morbid caufe, in which the difeafe may be faid to
confift. This only we know with certainty, that
where the febrile motions are violent and continual,

the difeafe haftens to a termination; where they are languid and feeble, or fuffer long interruptions, its duration is often drawn out to an undetermined length of time. Thus continued fevers, with inflammatory diathefis and much vifcular excitement, for the moft part terminate decidedly in feven or nine days; while thofe with low and languid motions, with long and diftinct intermiffions, as the quartan, and even fometime the tertian, continue for months, and decline at laft by flow and almoft imperceptible degrees.

It may feem that I have treated very fully of the critical days of fevers; yet before leaving the fubject altogether, there is one thing ftill which requires to be mentioned;—I mean the great proportion of fatal terminations, which happen on the even days. The even days were obferved to be fatal in the proportion of three to one, in thofe fevers, which came under my care during the time that I lived in Jamaica. The fact, which is curious and hitherto I believe unnoticed, was difcovered in the following manner. That I might the better trace the progrefs of nature through the whole courfe of the fever, a fubject which then engrofled my chief attention, I vifited often, and fpent much of my time in the apartments of the fick. Among other things, I difcovered the manner in which death more ufually approached. The natural courfe of the paroxyfm appeared generally to be finifhed, or the action of the febrile caufe feemed actually to have ceafed. The lightning before death, as it is termed, which has been generally attributed to the laft efforts of dying nature was frequently feen to take place. This was even fometimes fo remarkable, as to give flattering hopes of a favourable crifis; yet in a fhort fpace of time, the powers of life begun to fail, and at laft were gradually extinguifhed, like an expiring taper. —The crifis, ftrictly fpeaking, happened on the odd days, equally the fame in thofe who died, as in thofe

who recovered; only I had inaccurately, accuſtomed
myſelf to refer the critical period to that moment,
were the ſigns of criſis were firſt perceived; in the
other, I had conſidered it as happening at the hour
of actual death. Thus it was obſerved in thoſe fevers which terminated fatally on the even days, that
the powers of life, though irrecoverably exhauſted,
were not totally extinguiſhed by the paroxyſm of the
odd day. This paroxyſm, in ſhort, ſeemed to decline after the uſual duration. It left the body, in
ſome meaſure, free from diſeaſe; but ſo completely
deranged in the vital functions, that the action of
living, though it often went on for a few hours, could
not be continued long. In this manner, the hour of
death was frequently protracted to the even day; yet
death happened ſometimes on the even days, from
another cauſe. The decline of the paroxyſm, which
in many caſes was hardly perceptible, in others was
very plain. The diſeaſe terminated; but a new one
recurring, after a ſhort interval, ſpeedily put a period to exiſtence. In the mild fever of Jamaica,
death uſually approached in the gradual manner I
have juſt deſcribed; yet in caſes of much violence
and malignity, the fatal termination was frequently
on an odd day. In ſuch caſes the patient died in the
height of the paroxyſm, carried off by convulſions,
apoplexy, or other accident.

Thoſe authors, who, ſince the time of Aſclepiades,
have denied the power of critical days in fevers, are
numerous; and many of them poſſeſs conſiderable
authority in the medical world. Their opinions,
however, cannot be conſidered as of great influence
in the preſent caſe, though they may aſſert, that they
never have obſerved the pre-eminence of any particular days in terminating febrile diſeaſes; ſuch an aſ-
ſertion means but little; unleſs its author convinces
us, that he has adopted a method of inveſtigation by
which thoſe regular movements, if they actually ex-

ifted, could not fail to be difcovered. Truth in the prefent cafe, can only be known from minute and careful obfervation; but a train of minute obfervation is not likely to be the work of a bufy phyfician; and one, who is little employed, has not fufficient materials in his practice to engage his attention to a continued purfuit. I confider it as my own good fortune, to have been placed between the two extremes of idlenefs and too much bufinefs. In the country where I refided for fome time, the movements of nature were generally fo diftinct, as to be obferved without much difficulty; my practice likewife was fufficient to employ my mind, and not more than it could comprehend eafily; fo that I had fufficient leifure to write down, and to digeft the obfervations which I have related above. They afford, if I miftake not, fome facts which are precife and pointed; and which fuperfede a multitude of arguments. I will not venture to fay, that they remove all the myftery from this dark fubject; but I cannot help flattering myfelf, that they point out a road by which we may continue our inveftigations with fuccefs. The fubject of critical days is of fuch importance, as to demand every attention. A knowledge of it gives credibility to our art; whilft ignorance in this refpect is the fource of perpetual miftake and difappointment. There are many phyficians of the prefent day, who treat the idea of critical days with ridicule ; but their affertions only afford an argument of their own precipitancy, and fuperficial obfervation. The man in reality, who pretends to cure a fever, without a knowledge of the critical periods of nature, is no lefs prefumptuous, than the mariner, who undertakes to conduct a veffel through the ocean, without being inftructed in the manner of calculating her courfe.

# CHAP. IV.

THE general remote caufes of intermitting and remitting fevers have been fo fully invefti-gated by feveral eminent writers, particularly by the induftrious and learned Lancifi, that little remains to be added : nor perhaps fhould I have thought it ne-ceffary, even to have mentioned the fubject, were it not to take notice of fome opinions of the late Sir John Pringle, which appear to have been formed too precipitately ; and which, I can affirm from experi-ence, have been pernicious to the health of thoufands. It would be a very needlefs oftentation to adduce the authority of the ancients, to prove the general fource of the difeafe which is the fubject of the prefent trea-tife. The hiftorians, no lefs than the phyficians of every age, do not entertain a doubt, that fevers of the intermitting and remitting kind, owe their origin to exhalations from fwampy and moift grounds. Daily experience ftill proves it ; and there are few men whofe obfervations are fo circumfcribed, as not to know, that it is in the neighbourhood of fwamps, and near the banks of frefh water rivers, that thofe diforders chiefly prevail. But though it is only in the above fituations, that intermitting and remitting fevers are more peculiarly epidemic ; yet it likewife deferves to be remarked, that, independent of the particular circumftances of foil and local fituation, the endemic of champaign countries is fubject, in a greater or lefs degree, to an appearance of period-ical revolution. Mud and ftagnant water, in every climate, poffefs the materials of the caufe of this fpecies of difeafe ; but a combination of other cir-

F

cumftances is required to give them activity. Among
the principal of thofe circumftances, which call forth
this action, we may reckon the influence of a powerful fun. Hence, (as is commonly known), fome
fituations, which, in the colder months of winter,
are diftinguifhed for no particular difeafe, in the hot
months of fummer and autumn, are obferved to be
moft malignantly unhealthful.

The nature of this exhalation or caufe of fever,
though it has long been a fubject of enquiry, remains
ftill unknown. We plainly perceive it to be of various degrees of force, and in various ftates of concentration; and we can eafily conceive it to be varioufly modified and combined;—but we go no farther. It has been faid, to poffefs a feptic principle;
but this alone will fcarcely be thought fufficient, to
account for the very peculiar manner in which it
affects the human race. Some other quality is neceffarily joined with it, which our fenfes cannot lay
hold of. But though the ingenuity of man has not
hitherto been able to penetrate the intimate nature
of this caufe of fever, we ftill have it in our power,
in fome degree, to trace its effects on the human conftitution. We plainly perceive that an habitual expofure to it, is peculiarly *unfriendly* to the principle
of life, and in a very remarkable manner fhortens
the period of exiftence. In proof of this I mention
from good authority, that white females, born and
conftantly refiding in the lower diftricts of the province of Georgia in America, have feldom been obferved to live beyond the age of forty. Males,
fometimes approach near to fifty; while Europeans,
who had arrived at manhood before they came to the
country, often attain a good old age. The fact is
curious, and fhews, in a ftrong point of view, the
deleterious quality of the air of thofe climates. But
though the general nature of the country, which I
have juft now mentioned, is unhealthy in a high de-

gree; yet there are fituations, in the Carolinas and Virginia, which are deftructive of life in a ftill more remarkable manner. There is not on record, I am credibly informed, an inftance of a perfon born at Peterfborough in Virginia, and conftantly refiding in the fame place, who has lived to the age of twenty-one. When the Britifh army marched through this province, in the year 1781, I had the opportunity of feeing a native of this town, who was then in his twentieth year; but he was faid to be the firft, who had ever attained fo advanced an age. He was de-crepid, as if from the defects of time, and it did not appear that he could furvive many months. Yet it is not a little curious, that this man had never been much confined with ficknefs. The refiding con-ftantly in the fame pernicious air, feemed alone to have been fufficient fo remarkably to accelerate de-crepitude. But though the inftances I have men-tioned, afford fufficient proof, that this miafma is un-friendly to the principle of life; yet we are by no means inftructed, as to the manner, by which it be-comes fo. This feems to be one of the arcana of nature; and it will profit little to profecute it farther by conjecture. It will, however be an object of utility to mark the foils and fituations in which the exhala-tion moft abounds, and to trace the caufes which heighten or lower its activity.

The hiftory of the remote caufes of intermitting and remitting fevers, with all the circumftances con-nected with them, having been, as I faid before, fo fully inveftigated by others, I fhall only add a few curfory remarks, where the information does not feem to be fufficiently precife, or where the conclu-fions, which have been made, are not juftifiable by experience. It is an opinion, which, though it did not originate with Sylvius de le Boe, evidently gained weight from his authority—that a mixture of falt with frefh water, as corrupting more eafily, af-

fords a more noxious exhalation than frefh water
alone. Lancifi has mentioned the obfervation ; and
Sir John Pringle confiders it as an eftablifhed fact;
but the evidence, by which he attempts to fupport
his opinion, is not decifive. It would be in vain to
deny, that the neighbourhood of lakes or rivers, with
a mixture of falt water, is often highly unhealthful ;
yet we may affirm with confidence, that it is feldom
more fo, than where the lakes and rivers are perfectly
unmixed. In proof of this affertion, I might adduce
the example of Savanna la Mar in Jamaica, or draw
inftances from the numerous iflands on the coaft of
the Carolinas ; where fea and river water are often
blended together in various proportions ; to which
might be added, the more particular evidence of the
relative healthinefs of the banks of rivers. So far
as I have obferved, the ufual endemic was lefs frequent, and lefs formidable on the banks of rivers,
after their waters became mixed with thofe of the
fea, than before this happened ; unlefs the circumftances were in other refpects more favourable for
the production of the difeafe. Hence there is but
little reafon for fuppofing, that there actually exifts
any degree of mixture of falt with frefh water, at
leaft of running water, which abfolutely heightens
the noxious quality of the exhalation. The above
is an opinion of fufficient confequence to demand inveftigation : but there is another advanced by this
celebrated author, worfe founded, and of ftill greater
concern, which I fhall likewife mention. From an
idea that a free circulation of air, is of all things the
moft effential to the prefervation of health, Sir John
Pringle enjoins in a very pofitive manner, not only
that open ground, but that the banks of large rivers
fhould be chofen, in preference to other fituations, for
the encampment of troops. This author's opportu-
tunities of information were good ; his opinion has
therefore gained weight, and his advice, I am afraid,

has been often fatally followed. It would be no difficult tafk to produce teftimonies, from both ancient
and modern hiftory, of the unhealthinefs of thofe fituations, which Sir John Pringle has thought proper
to recommend; but at prefent I fhall confine myfelf
to that, which has more immediately fallen under my
own obfervation. The inftance I fhall mention, is
only a fingle one; but it proves fo clearly the danger
of encamping on the banks of frefh water rivers, as
to render all others fuperfluous. In June 1780, the
firft battalion of the 71ft regiment was detached to
the Cheraws, where it encamped on open ground,
within five hundred paces of the river Pedee. The
people of the country, taught by experience, fuggefted
the propriety of drawing back the encampment into
what is called the Pine-barren, affigning as the caufe
of their advice, that the diftance, as well as the cover
of the wood, might be a fecurity againft the damps
of the river, which were obferved to be extremely
noxious in that climate. A pofition in wood, acceffible on all fides, would not perhaps have been military; fo that no alteration was made. The other battalion of the regiment joined in July. It arrived in
perfect health, and encamped likewife on open ground;
but ftill nearer the river. In a fortnight the inter—
mitting fever began to make its appearance; and in
lefs than three weeks, more than two thirds of the
men were ill; whilft fcarcely one of the officers had
efcaped. The officers, it muft be remarked, encamped in the rear of the men, and immediately on
the bank of the river, the courfe of which was uncommonly flow at this place; while its banks, though
high, were oozy and foul. There are few inftances
on record perhaps, where a degree of ficknefs,
greater than the prefent, has been obferved in fo
fhort a fpace of time. The firft battalion, however,
did not fuffer in the fame proportion. The ground

F 3

of encampment was not only at a greater diftance
from the river; but being alfo nearer to a wood,
many of thofe, who were not confined by their duty
to a particular fpot, found a convenient fhelter in its
fhade, from the powerful heat of the fun.   Thefe I
muft not omit to mention, were the leaft fickly of
the whole encampment.   The above is an important
fact.   It proves clearly, that no ideal circulation of
air can counterbalance the noxious exhalations from
rivers; and it likewife affords a prefumption, that inftead of danger, there is fafety in the fhelter of wood.
But with regard to this, no abfolute rule can be given.
It muft generally be decided by local circumftances,
whether wood, or open ground are to be preferred
for the encampment of troops.   Upon the whole,
however, there are many reafons to induce us to believe, that as an encampment is not only more military in the body of a wood, than in open ground furrounded by woods; fo it is likewife more fafe with
refpect to health ; particularly if within the reach of
effluvia from fwamps or rivers.   The reafon which
offers is obvious.   The wood not only ftops the progrefs of noxious vapours carried from a diftance ;
but it alfo covers the earth from the immediate action of the fun—the powerful caufe of exhalation;
in doing which, it perhaps, does more than counterbalance the lefs free circulation of air, or the greater
dampnefs of the ground.   But left the authority I
have mentioned, fhould not be thought fufficient, the
opinion receives farther confirmation from the teftimony of the ancients.   Hiftories abound with examples of deftructive epidemics, which have followed
the cutting down of groves, which covered mo-
rafles, or which intercepted the progrefs of marfh exhalation.   America alfo furnifhes daily inftances of
a fimilar truth.   In this country the unhealthinefs of
a place is often obvioufly increafed, by cutting down
the woods of the neighbouring fwamps : hence no

rule is more liable to exceptions, than that which has been fo generally enforced; viz. that clearing a country of its woods invariably renders it healthy: unlefs the grounds be drained and cultivated, as well as cleared, the effect is likely to be the reverfe.

It would be curious and ufeful, could we trace this miafma or caufe of fever in its progrefs. I do not deny that the noxious exhalation may be accidentally enveloped in fogs; but it is not neceffarily fo; and I add, that the dews of night, unlefs as an exciting caufe, are lefs pernicious than has generally been imagined. Low grounds, in the fame manner, are not always unhealthy; as high and dry fituations fometimes afford no protection againft the ravages of this difeafe. The fituation of the encampment which the 71ft regiment occupied at King's-bridge, in the year 1778, affords a curious and direct proof of the truth of this opinion. About two hundred paces to the right of the fpot, on which the tents were pitched, was a tract of low and fwampy ground; but the immediate fituation was dry, and of confiderable elevation. The right was *particularly* fo; yet it was principally on the right, where the difeafe raged with violence. The left, though on low ground, over which fogs frequently hung till late in the day, fuffered in a much fmaller proportion. From this we might infer, that a dry and elevated fituation is by no means exempted from intermitting and remitting fevers: but the great degree of ficknefs, which happened to thofe people, who not being confined by the nature of their duty to one particular fpot, pitched their tents on a hill in the rear of the encampment, proves it clearly. The ground, which thofe perfons made choice of was directly in the tract of air, which blew over the fwamp. It was dry and fcarcely ever covered with fogs; yet there was not an individual among them who encamped upon it, who did not fuffer from this raging epidemic. The prefent inſtance, with many others which I might adduce, leaves little room to doubt, that inſtead of expoſing encampments to ſtreams of air, which blow from rivers or ſwamps, it ought to be our principal buſineſs to guard againſt thoſe noxious effluvia, by the interpoſition of woods or riſing grounds. Exhalations which are the cauſes of fevers are ſubtile, and ſeem to be pernicious, chiefly in their aſcent :—viſible damps or night dews are comparatively innocent.

So great is the importance of preſerving the health of an army in the field, that the choice of encampments ought to be made a ſubject of particular enquiry. The opinion of Sir John Pringle on this head, (which, in fact, is an opinion of theory rather than obſervation), has been followed too long without examination. The directions of this author are influenced wholly by the dread he entertained of a contagious or hoſpital fever; but a contagious fever, is ſeldom a diſeaſe of the field; and has, perhaps, ſcarcely ever been known to make its appearance in a moving camp. Diſeaſes of the field are often epidemic, ſometimes malignant, but rarely contagious. I even doubt if the dyſentery, whilſt a camp-diſeaſe, is ſo in any remarkable degree. It was not ſo at leaſt in America, in thoſe campaigns, where I had the opportunity of knowing the ſtate of the army.

The general remote cauſe of intermitting and remitting fevers, conſiſts, as was mentioned before, in inviſible exhalations floating in the air. Theſe are more copious in ſome ſituations than in others; and appear to be rendered more or leſs active by a great variety of cauſes. Among the number of thoſe cauſes which have been accuſed of exciting fever, it has been uſual to reckon exceſs in drinking. It cannot be denied, that this cauſe, in ſeveral caſes, has brought forth the diſeaſe, when it probably would not have otherwiſe appeared; yet it has been likewiſe obſerved that a debauch of wine has ſometimes reſtored the

body to health, when languifhing under the influence
of this diforder in an obfcure or irregular form. The
moderate ufe of wine, however, has been generally
recommended as a prefervative in times of great
heat, and epidemic ficknefs :—and under limitations
it undoubtedly is of ufe. In a time of very preffing
calamity, the oracle of Delphi gave its fanction to the
prefcription, and hiftory bears teftimony to its fuccefs. But befides excefs in drinking, cold and fatigue
have likewife been confidered among exciting caufes
of fever. In fhort, whatever exhaufts or diminifhes
the activity of the powers of life, may be juftly
viewed in this light. Yet ftill I muft obferve that
neither cold, fatigue, nor any of the caufes of this
train, give occafion to a proper intermitting or remitting fever, unlefs the predifpofition to the difeafe
be particularly ftrong. As a proof of this, I muft
beg leave to mention a fact, which fell under my
own obfervation. In an expedition into South Carolina, in the year 1779, a part of the army was near
five hours in paffing Purifburg fwamp. The men
were always up to the middle, fometimes up to the
neck in water. The cold and fatigue were both
very great, and a fit of intermitting fever was the
confequence in a great number of the foldiers: yet ·
it was only in a few inftances that the difeafe went
through a regular courfe, though there was even a
general pre-difpofition to it, in the habits of almoft
all the men who compofed the detachment. The
moft of them had fuffered from it feverely the preceding autumn; and a temporary return of it, was
generally obferved to follow any extraordinary exertion, or the application of a debilitating caufe. The
above caufes are generally reckoned exciting caufes
of fever ; but befides thefe there are feveral others
of confiderable power, which as being commonly
known, I fhall not now fpend time in enumerating.
There however ftill remains one, which, though

vèry univerfal, and perhaps more powerful than any other, has hitherto been little attended to. The approach to the new and full moon, in fome degree, perhaps in every part of the globe, but particularly in the Weft-Indies, appears to be connected with the invafion and relapfe of fevers, in a very remarkable manner. This obfervation has been hinted obfcurely by one or two authors; the idea has been treated with ridicule by others : and it muft be confeffed, that the facts, which have hitherto been produced in fupport of the opinion, are extremely vague and equivocal. I fhall therefore enter a little more minutely into the fubject, and ftate circumftantially the evidence, from which I have been led to confider the approach to new and full moon, as a powerful exciting caufe of fever.

That the moon exerts fome influence on the human frame, and that her different appearances are more or lefs connected with the progrefs and iffue of difeafes, does not feem to have altogether efcaped the notice of the ancients. In a fragment of Hippocrates, in the edition of Vander Linden, we find a detail of the different afpects of the moon and planets, with their combined influence on the fate of difeafes ; but the ftyle and manner of this little tract are fo perplexed, that I do not pretend to underftand its meaning. Galen had likewife fome obfcure ideas on the fubject; but he has left us nothing clear and explicit. The Arabian writers are alfo confufed and inaccurate, fo that the firft circumftantial evidence of the influence, or connexion of the moon with the human body, is found in the works of Ballonius, a French phyfician of the fixteenth century. The fact which this author records, though not altogether in point, is curious. A Parifian lady of quality appears by the account of Ballonius, to have been very fingularly affected during an eclipfe of the fun. Her complaint threatened nothing dangerous, and her phyficians

were amufing themfelves with obferving the progrefs
of the eclipfe, when they were fuddenly fummoned to
her affiftance. In the moment when the eclipfe was
deepeft, fhe had the appearance of dying ; but thefe
threatening fymptoms decreafed with the decreafe of
the eclipfe ; fo that fhe at laft returned to her former
ftate. This is only a folitary inftance, and perhaps
might be reckoned accidental. We may however
add to it the general teftimony of Ramazzini, who
lived at Modena in the beginning of the prefent century. This author's obfervations, indeed, are by no
means precife ; yet he was convinced by them, that
the courfe of Epidemics was confiderably influenced
by the particular ftate of the moon. It is almoft
needlefs to mention Dr. Mead, who wrote a treatife
exprefsly on the moon's power on the human body.
The facts which this writer has collected, afford a
reafonable prefumption, that this planet is not without fome influence in feveral difeafes to which man
is liable ; but we find not any thing in the work,
which patricularly relates to fevers. I fhall mention a
fact recorded by Dr. Grainger. It is the moft circumftantial I have yet met with ; and the ftrongeft to be
found perhaps in the writings of any European phyfician. Dr. Grainger, who was a furgeon of the
army, ferved in the Netherlands about the years 1746
and 47, and wrote a treatife on the intermitting fevers of that country. Among other obfervations he
takes notice of a circumftance which occurred to him
at that time, and which he then confidered as fingularly curious ; viz. that twenty of the men of the
regiment, of which he had the charge, were feized
with this fever, which was then epidemic, on the day of
a folar eclipfe. He has not made any application of the
fact. It furnifhes however a very fubftantial evidence, of the influence or connexion of this planet
with the invafion of febrile difeafes.

It appears to have been long known in India, that

fevers have a tendency to relapfe about the new and full moon, and particularly at the time of eclipfes but Dr. Lind of Windfor is the firft, who brought the knowledge of the faƈt to Europe. In an inaugural differtation, publifhed at Edinburgh (I do not exaƈtly recolleƈt the year), this author obferves, that this opinion prevailed very generally in the Eaft. He adds likewife, that fome inftances occurred in his own praƈtice, which gave him caufe to believe that the faƈt was well founded. Dr. Lind continued of this way of thinking for feveral years after his return to England. He does not indeed at prefent deny the faƈt. He only fuggefts that it may admit of a different explanation, from that which he had given in his firft publication. The fpring tides, as they overflow the low grounds, according to his prefent opinion, afford a more probable caufe of the uncommon increafe of fevers about the new and full moon, than the direƈt influence of the planet itfelf. I will take the liberty however to add, that this opinion has been offered to the public, from a very imperfeƈt view of the fubjeƈt. I can affirm, even from the confined circle of my own experience, that a connexion, between the moon and the invafion of fevers, certainly takes place in diftriƈts remote from the fea; and I believe it is generally known, that a fever, or the paroxyfm of fever, is not commonly the inftantaneous confequence of expofure to its remote caufe; which ought to be the cafe, if this author's reafoning were juft.

The next, and indeed the only author who has written profeffedly on the influence of the moon in fevers, is Dr. Balfour; a gentleman who refided feveral years in India, and who praƈtifed with reputation in the fervice of the Company. This author pretends to have inveftigated the fubjeƈt with care and attention; but there appears in reality, to be more theory and general affertion in the treatife than circumftantial fact. The refult of his obfervations he informs us, amounts to this: viz. that the three days which precede, and the three days which follow new and full moon, are remarkable for the invafion and relapfe of fevers; that the day of the full moon, and the day of the change of the moon, are the moft remarkable of all: and farther, that the days which follow, are, in general, more remarkable than thofe which precede.

I have now brought together the fubftance of what is found in the writings of thofe authors who have mentioned curforily, or treated profeffedly of this fubject. There is not in any part of it, if we except the inftance recorded by Dr. Grainger, any thing accurate and precife enough to enable us to form an opinion. What has fallen under my own obfervation, I would flatter myfelf, is lefs ambiguous; and though it may not be fo explicit, perhaps, as to eftablifh the doctrine completely, it may at leaft affift us, I hope, in approaching nearer to the truth.— I fhall relate it in a few words.

When I arrived in Jamaica, in the year 1774, I had no other knowledge of the influence of the moon in fevers, than what I retained from a curfory reading of Dr. Lind's differtation. I remember, however, to have mentioned the circumftance to feveral practitioners, who had lived many years in the ifland. As I conceived there was a fimilarity between the climates of Jamiaca and Bengal, I thought it not improbable, that fome of the practitioners of the country in which I then was, might fupply me with fatisfactory information on the fubject. There were none of them, however, who acknowledged that they had ever obferved any connection between the moon and febrile difeafes; neither were there many of them, who feemed difpofed to give credit to its exiftence. Twelve months or more elapfed without my having paid any further regard to the fact, when an acciden-

G

tal relapfe of fever, happening near the time of full
moon, recalled Dr. Lind's obfervation to my memory. It likewife brought to mind a circumftance,
which till then I had overlooked. I had feen frequently, though without attending to it particularly,
that three or four of the foldiers of a company of the
60th regiment, who were quartered at Savanna la
Mar, and of whom I had the care, were attacked with
fever on the fame day; whilft it feldom happened, that
any other febrile illnefs made its appearance in the
garrifon, for the enfuing fortnight. This having
been obferved oftener than once, at the time the
moon was near full, a hint fuggefted itfelf, that the
caufe, which was faid to influence relapfes in India,
might here have an effect on the original invafion.
But in order to afcertain the truth of this conjecture,
which I confidered as a matter of fome importance,
I provided myfelf with the almanack of the year 1776,
and marked, in the blank leaf of it, the precife date
of attack, of all thofe fevers which came under my
care. In looking over thofe memoranda at the end
of the year, I found I had put down thirty cafes of
proper remitting fever, the invafion of twenty-eight
of which was on one or other of the feven days, immediately preceding new or full moon; that is in the
fecond and laft quarters. The fame plan of obfervation was continued through the following year,
and the refult, though not exactly the fame, was
fimilar. Of twenty-eight cafes, which where found
in the almanack, twenty-two were in the periods
above-mentioned: that is in the fecond and laft quarters of the moon. It deferves however to be remarked, that three of thofe fix cafes, which were
not in the common period of invafion, happened actually on the day of new moon;—a few hours after
the change had taken place. But befides thofe cafes
of proper remitting fever which I have mentioned,

there were likewife found in the almanack many days fevers and flight feverifh diforders, the invafion of the greateft number of which was likewife in the ufual period.

The above is a literal ftate of the cafe as it food in the almanack :—fome remarks and obfervations, however, were added, of which the following are the principal : viz. That, though the whole of the fecond and laft quarters of, the moon is included in this period of invafion ; yet the four days immediately preceding new and full moon, were more particularly diftinguifhed for thofe febrile attacks : that in the dry feafon, which is reckoned the moft healthy, the time of invafion was more clofely connected with the new and full moon, than in the wet and fickly months, particularly when the ficknefs was epidemic, or of a bad kind : and laftly, that this influence, or connexion was more apparent in the foldiers of the garrifon, who were expofed to few occafions of difeafe, excefs in drinking excepted, than in the inhabitants of the town and country, whofe occupations carried them oftner to places of unhealthy fituation ; or whofe modes of life obliged them to fubmit to more various hardfhips or to greater fatigues than fell to the lot of a foldier in times of peace.

I fhall further beg leave to add, that I went to join the army in America, in the year 1778 ; and that I continued in that country, the train of obfervation on this fubject, which I had begun in the Weft Indies. The regiment, in which I ferved, was encamped during the months of June and July on a healthy part of York-ifland. Fevers were rare ; and the time of invafion, of fuch as did appear, was chiefly confined to the fecond and laft quarters of the moon. In the beginning of Auguft, the encampment was removed to King's-bridge, where it occupied a very unhealthy fituation. The intermitting fever foon made its appearance. It extended in fome degree to the whole battalion; but raged with particular violence on the right, which bordered on low and fwampy ground. The approach to new and full moon never failed, even in this climate, to increafe the number of the fick; yet it deferves to be remarked, that this increafe was always fmaller in proportion, in that part of the battalion, which lay contiguous to the fwamp, where the difeafe was highly epidemic, than in the other extremity of the encampment, where it prevailed in a lefs degree. But ftill upon the whole, when the regiment moved from their ground, in the beginning of November, of a hundred cafes of intermitting fever, which were marked in the almanack, eighty were found to have commenced in the ufual period of invafion; that is, in the fecond and laft quarters of the moon. It is fomewhat remarkable, that relapfes were in a fmaller proportion. This regiment, fome parts of the medical hiftory of which I defcribe, embarked on an expedition for the fouthward in November, and arrived at its deftination in Georgia, in the latter end of the year. It remained in the fouthern provinces, and ferved every campaign till the capitulation at York-town. The fame train of obfervation was continued during this intervening fpace, and the fame influence of the moon feemed in general to prevail; but the notes having been loft, I cannot now exactly afcertain the degree in which this influence took place. Of this, however, I am certain, that even in times of the greateft epidemic ficknefs, when the connexion was evidently weakeft, the number of the fick was generally doubled in the periods approaching to new or full moon.

We cannot avoid concluding, from the facts which I have ftated above, that the approach to new and full moon, or fomething connected with that approach, may be juftly confidered as a powerful exciting caufe of fever. The circumſtances, indeed, which I have mentioned, are ſo clear and unequivocal as to leave little room for doubt: nor did I entertain any, till I found that the obſervations of Dr. Balfour, on this ſubject, were ſo ſtrikingly different from mine. Bengal and Jamaica are diſtant from each other; yet few people will be diſpoſed to believe, that ſo great modification of a general cauſe has ariſen ſolely from this diverſity of climate. Dr. Balfour muſt ſpeak for himſelf. For my own part, I can only ſay, that what I ſaw I have related with truth. As I have told the manner in which the idea aroſe, with the manner in which the inveſtigation was conducted, I leave the concluſion to be formed by the reader.

G 3

# CHAP. V.

THEORIES of the proximate caufe of fevers, or more properly modifications of theories, are fo numerous, that a whole volume would fcarcely be fufficient to give any tolerable account of them. It is a tafk indeed which I fhall not undertake; yet I hope it will not be altogether fuperfluous, to give a curfory view of the principles, which have directed the conjectures on this fubject in different ages. The principles are, in fact, fewer in number than at firft fight they appear to be. Phyficians, ambitious of raifing their name and reputation, have fhown great induftry in multiplying and modifying opinions; yet it does not appear, that they have produced any great variety of theories, which are fundamentally diftinct.

The ancients, who were little acquainted with chemical principles, or with the qualities and properties of the nervous fyftem, placed the proximate caufe of fevers in fome fignal fymptom of the difeafe, fuch as increafed heat, or abounding bile; or entering ftill farther into the fields of fpeculation, ventured to attribute it to derangements in the permeable canals of the body, or to affections of the humours, or circulating mafs of fluids. Hence obftruction of pores, plethora, error loci, lentor and vifcidity, or putrefaction of the humours, have all feverally, at different times, or by different authors, been confidered as the immediate or proximate caufes of this difeafe. The theories, which prevailed in the fchools till the beginning of the fixteenth century, did not often extend farther than to the caufes which I have mentioned: but after that period, the difcoveries of

the famous Paracelfus opened a road to innovation
in medical reafoning. The followers of this author,
if not numerous, were enthufiaftic and vociferous.
They indulged in the wildeft extravagance of conjecture ; and their opinions, for a confiderable time,
were combated with the authority of Galen, rather
than with folid argument and accurate reafoning.
At laft the difputes between Chemifts and Galenifts
beginning to fubfide, the chemical theories became
incorporated with the doctrines of the mechanic philofophy, which were revived more than a century
ago, and which ftill maintain fome influence in the
common fyftems of phyfic. In the mean time happened the important difcovery of the circulation of the
blood ; but no immediate change, in the manner of accounting for fevers, enfued immediately in confequence
of it. Yet as from this period the refearches of phy-
cians began to be conducted on a more extended
plan, fome parts of the fyftem were brought into
view, which had been formerly little attended to.
The nervous fyftem, which had been in a manner
overlooked for many ages, was now found to be of
importance in the economy of the animal machine ;
and authors foon began to confider it, as affording a
probable feat for the proximate caufe of fevers.
Among the firft of thofe authors, who viewed it in
this light, we reckon Borelli and Dr. Cole ; the one
of whom in Italy, the other in England, propofed
much about the fame time, new and different opinions about the proximate caufe of fevers. Their
conjectures, I muft confefs, are far from being probable ;—(that of the Italian is fcarcely ingenious)
yet they deferve to be mentioned in this place, as
being among the firft attempts to bring into view a
part of the fyftem, which is very effential in enabling
us to account for many appearances in febrile difeafes.
It is commonly believed, that the nervous fyftem
was not difcovered to be a part of material import-

ance; either in the functions of health, or in the affections of ficknefs, till the laft century. This, in fact was generally the cafe; yet I muft not omit to mention, that we meet with an expreffion in the writings of Hippocrates, viz. τα ερμωντα η νορμωντα σωματα, which might incline us to be of opinion that this phyfician was not altogether ignorant of the influence of the nervous power; and that he actually confidered this principle of the conftitution to be of much importance in the management and cure of difeafes. After Hippocrates, Van Helmont, under the whimfical appellation of Archeus, afferted more directly the dominion of the fentient principle. He has indeed applied its operations more particularly to affift him in explaining the theory of fevers; but it has been a misfortune that the opinions of this author have been generally lefs attended to, than perhaps they deferve: fo that it has been cuftomary to confider, the celebrated Hoffman as the firft, who fuggefted the idea, that the proximate caufe of fever depends on a derangement or affection of the nervous fyftem; at leaft he is the firft, who delivered a fyftem on the fubject, which can in any degree be confidered as rational and confiftent.

It will not be an eafy tafk, to give a clear and diftinct view of that, which has been confidered by the ancients, as the proximate or immediate caufe of fevers. The language of the earlieft writers is not by any means precife in this refpect; and we fhall frequently, perhaps, have difficulty from the ambiguity of expreffion, to diftinguifh from each other the definition, the remote and occafional, or the immediate and proximate caufe of the difeafe. The proximate caufe of a difeafe, it muft be remembered, is a caufe which conftantly and uniformly produces its refpective complaint; and without which this complaint cannot even for a moment exift. It is, in fhort, the firft effential derangement, which the ac-

tion of this caufe produces in the frame of the fuf-
ferer: but though we know this to be certainly true,
yet we have made no progrefs in difcovering the nature of this derangement. The firft action of the
caufe of fever is obfcure, and fome part of the derangement which it occafions, has hitherto probably
pafled over unnoticed, even by the moft accurate
obfervers.

If we attempt to give a view of the fucceffive
conjectures, which, at different times, have been offered to the public concerning the proximate caufe
of fevers, it will be neceffary to begin with Hippocrates. We may collect very clearly from the
writings of this author, that an increafe of the heat
of the body had afforded, to the ftill more ancient
phyficians, the firft idea of the eflence or immediate
caufe of fevers. This feems to have been the idea
of the moft ancient profeffors of medicine. Hippocrates in fome degree fubfcribed to it; yet this author
feems likewife to doubt, if the fimple increafe of
heat alone is fufficient to conftitute a proper fever,
or that it can with propriety be confidered as the effential proximate caufe of the difeafe. But though
Hippocrates raifes this objection to the common opinion concerning heat, yet he ftill leaves us in doubt
with regard to the opinion which we ought to adopt.
His ideas are fluctuating and uncertain. We find in
the different parts of his works, obftruction, plethora, miafmata or bile, all feparately confidered, as
immediate caufes of fever. But fuch caufes, I may
add, where they do take place, are in fact only more
remote or diftant caufes. Neither miafmata, bile,
nor obftruction, are circumftances on which the exiftence of fever invariably and neceffarily depends;
at leaft fuch caufes require to be in a certain ftate
of modification, which is yet undefined, before they
are capable of actually producing the difeafe. Bile
bears a very confpicuous part in the Hippocratic

doctrine of fevers. The fabric, indeed, which our author raises on this principle, is fanciful, and in many respects, ill founded; yet, as modified by the fertile genius of Galen, it passed on through a succession of many ages: nor is it, even now, altogether banished from the language of practitioners.

Such are the hints concerning the causes of fevers, which I have been able to collect from the writings of Hippocrates. The expressions are often obscure or equivocal; and we can scarcely say, that an opinion can be formed from them which deserves the name of a theory. The successors of this great physician were, perhaps, too sensible of this defect; and therefore attempted to fabricate other opinions, which might be more explicit and distinct. Among the first of those attempts, we may reckon the hypothesis of Diocles of Caryftus, a physician who lived at an early period, and who was highly esteemed by his contemporaries. Fever, according to this author, is not so much a primary disease, as a symptom of some other affection. Wounds, tumours, and many other accidental causes, have certainly been observed to give rise to symptoms which have been usually denominated fever; yet neither wounds nor inflammations have been generally observed to give rise to a proper fever. I will not however deny, that wounds, or inflammations, occasionally prove exciting causes of proper fever, where there is a strong disposition to the disease, existing in the constitution, at the time those accidents have happened. It does not appear that this theory of Diocles gained much ground with succeeding writers; yet it was, perhaps, the cause of introducing the distinction of primary and symptomatic into the history of fevers; a distinction, which is frequently of consequence in practice. But I must further add, that though the opinion of Diocles is not admissible in its literal meaning; yet, in a modified sense, it is not altogether without foundation.

The fymptoms of fevers are undoubtedly indications of a derangement of the body from its healthy ftate; but when we have faid this,' we can fay no more.— The nature of the derangement, which in its firft beginnings is not obvious to the fenfes, neither the ancients, nor the writers of the prefent age have, as yet, been able to afcertain.

Not very long after Diocles, Erafiftratus, a native of the ifland of Cea', and phyfician at the court of Antigonus, furnifhed a conjecture concerning the caufe of fevers, which is mentioned both by Celfus and Galen, and which appears to have originated in his anatomical refearches. As Erafiftratus directed his purfuits particularly to the fanguiferous fyftem: fo impreffed, perhaps, with an idea of the importance of that part of the body on which his thoughts had been chiefly employed, he ventures to hazard the opinion, That the immediate caufe of fever depends on a certain error loci, or transfufion of the red blood into the arterial channels: and this, he moreover adds, proceeds from repletion.—The opinion originates from an anatomical error, and on that account need not detain us any longer.

The next author, of whofe opinion on this fubject any diftinct traces have been tranfmitted to us, is Afclepiades, the Bythinian, a man who feldom treats the doctrines of his predeceffors with refpect. In his rage for innovation, Afclepiades attempted to change or modify the theories of thofe who had gone before him, in fuch manner, as to hope to impofe a conjecture on the world, which might, at leaft poffefs fome exterior claims of novelty. He allows with the moft ancient phyficians, that the infeparable fign of fever, or its effential part, confifts in an excefs of heat; but having adopted the doctrine of atoms, which was conveyed to the Greeks by Democritus of Abdera, he pretends to account for the difference of types by a difference in the fize of the corpufcles, which he fuppofes to be formed by a combination of indivifible atoms. Thus we fee that obftruction in the permeable canals of the body, in this writer's opinion, conftitutes the theory of the proximate caufe of fever: on which principle we may likewife conclude, that the modern doctrine of lentor and vifcidity has built its foundation.

The author, whom I have laft mentioned, may actually be confidered as the original founder of the methodic fect. The principal tenets of this fect of phyficians have been tranfmitted to us by Celfus, Cælius Aurelianus, or Galen; but the doctrines, which they promulgated, have not been very fully and perfectly explained. The great divifion of Themifon, into ftrictum et laxum, furnifhes a very fimple view of difeafes. Fevers are included in the firft order of derangement; and in this refpect, may be confidered as depending on a caufe fimilar to the obftruction obfcurely hinted by Hippocrates, or more explicitly defcribed by Afclepiades. There is this difference, however, between thefe refpective opinions, that the earlieft writers feem to have referred the obftruction to fome change in the humours or circulating mafs; while the *methodics* appear to have attributed it more directly, to a change in the capacity of the containing veffels. Hence we may infer, without any improper latitude of interpretation, that the ftrictum of Themifon and Theffalus comprehends the fpafmodic conftruction of capillaries, which has lately made fo confpicuous a part in the theory of febrile difeafes. This theory of the methodics, where the nervous and fibrous fyftem have been more regarded than the humours, or circulating mafs of fluids, was principally followed at Rome, for more than a hundred years. At laft Galen, who was a very unqualified admirer of Hippocrates, exerted himfelf fo fuccefsfully in reviving the humoral doctrine of his mafter, that the methodic fect began to fink rapidly into decay; and after a fhort time its traces were totally obliterated.

The frequent blanks, in-medical hiftory, make it no eafy tafk to give a connected view of the fluctuating fyftems of the ancient phyficians. The works of every writer of the methodic fect have perifhed, except thofe of Cælius Aurelianus: neither have we been able to difcover any new opinion, or modification of opinion, concerning the proximate caufe of fevers, between the time of Afclepiades or Themifon, and the great commentator of Hippocrates, except that of Athenæus. Athenæus, who was the head of the fect of Pneumatics, ftood high in efteem among his contemporaries and fucceffors. This author ventured to fuggeft a new hypothefis, or more properly perhaps, only extended, and more fully explained a doctrine, of which the obfcure traces may be difcovered at an earlier date. The general caufe of fever, in this writer's opinion, confifts in a putrefaction, or putrefcent ftate of the humours. Hippocrates feems to have entertained fome indiftinct idea of the fame kind; and thofe, who have been inclined to this way of thinking, both in ancient and in modern times, have neither been few in numbers, nor contemptible in authority.

Galen, who has written on moft parts of medical fcience more learnedly than his predeceffors, has difcuffed very fully the fubject of the proximate caufe of fevers. Amidft the luxuriance of this author's colouring, it is fometimes difficult to lay hold of the precife idea; at the fame time, that it is oftener tedious than inftructing to follow him through the maze of his fanciful and inconclufive reafonings. I fhall not therefore enter into a minute detail of his arguments; but ftill I conceive it may be ufeful, particularly to thofe who have not the opportunity of confulting his voluminous, and in fome refpects ill digefted works, if I comprefs into narrow compafs the lead-

H

ing principles of his general doctrines. In the firſt place, the opinion, hinted by Hippocrates and adopted by moſt of his fuccefſors, that the efſence of fever conſiſts in a certain derangement of heat, is exprefsly maintained by Galen, who explains more elaborately than his predeceſſors the various circumſtances, which influence or modify this general caufe of the difeafe. Galen aſſumes, indeed, as a fundamental poſition, that heat any how, or any where excited, communicated to the heart, and from the heart to the reſt of the body, conſtitutes a fever; yet he afterwards adds more explicitly, that a preternatural heat does not conſtitute a fever, unlefs it is communicated to the heart; which is confequently to be confidered as the principal feat and refidence of the febrile affection. Having, as he imagines, eſtabliſhed this fundamental principle, he proceeds to invcſtigate, more particularly, the parts of the body where the heat refides, and the caufes by which it is generated, propagated, or ſo modified, as to produce the difeafe in its different forms. But, that he may the better explain his meaning clearly, he divides fevers into three different kinds : viz. the hectic, or habitual, the humoural, and the ephemeral. The firſt he ſuppofes to arife from an affection of the folids, or containing parts; the ſecond from ſome derangement of the fluids, or contained partc; and the third from ſome diſturbance of the fpirits, or that part of the frame which we, perhaps, now diſtinguiſh by the name of nervous fyſtem. He adds in the next place, that putrefaction is the medium, by which fever is excited, where the fluids or humours are the fubject of the difeafe, contiguity and continuity, where the illnefs affects the habit or folid parts ; and where the effects are tranſitory aud fleeting, he attributes the caufe principally to the rapid movements of the fpirits, or nervous influence. And laſtly, he attempts to complete his theory, by explaining the different types of humoural fevers, on the fuppofition of a ftate of putrefaction in the different humours, from which he fuppofes the difeafe to arife. On this fubject he has deviated very materially from his mafter Hippocrates, though he probably drew his ideas from the hints, which are found in that author's works. Hippocrates explains, or attempts to explain the various types of fevers, by a fimple difference in the quantity of the bile. Galen, on the contrary, as we have faid juft now, endeavours to account for this phenomenon, by a fuppofition of putrefaction in the phlegmatic and bilious humours, which bear fo confpicuous a part in his theoretical fyftem. Thus Galen fuppofes, that a putrefcent tendency in the blood gives rife to a continued fever; a fimilar difpofition in the phlegm difpofes the difeafe to appear in a quotidian form: putrefaction of the yellow bile determines the type to be of the tertian kind; whilft a like tendency, in the black bile, regulates the movements of the quartan period.—It is unneceffary to make any remarks on the bafelefs fabric, which this author has offered to the world, concerning the proximate caufe of fevers.—Its inconfiftency and infufficiency are perfectly obvious.

After the time of Galen there does not appear to have been any material change, in the manner of accounting for fevers, for many ages. Aetius Amidenus indeed fuggefted fome reftrictions and explanations in certain fpecies of fever, which do not feem to have been fo explicitly marked by the commentator of Hippocrates. Inftead of confidering putrefaction as the fole means of exciting heat in every fpecies of humoural fever, Aetius ventures to infinuate, that there is no ftate of actual putrefaction in the ·ινοχος, or that fpecies of difeafe which is purely inflammatory, the caufe of which appears to be fimply an inordinate fermentation or ebullition of the blood. But except in this inftance, the fucceeding Greek phyficians do not feem to have departed, in the leaft, from

H 2

the direct footſteps of Galen. The Arabians likewiſe, among the principal of whom we may reckon Avicenna, adopted his general doctrines, and modes of reaſoning, only Avicenna defines more exprefsly than others had done before him, that fevers of all denominations ariſe immediately from a preternatural heat of the heart; in doing which, he ſeems to have extended the influence and power of that quality which preceding authors in looſer terms had conſidered as the general cauſe of febrile diſeaſes.

The doctrines of Galen, with ſome immaterial innovations of the Arabian phyſicians, wholly occupied the ſchools of medicine, till the beginning of the ſixteenth century, about which time Aureolus Philippus Theophraſtus, commonly known by the name of Paracelſus, effected a revolution of opinions, which marks an important period in the hiſtory of the medical art. Paracelſus, who was a man of a ſingular turn of mind, ſpent the earlier part of his life in travelling among the nations of Aſia; where he probably acquired ſome knowledge of chemiſtry, in which ſcience the Arabians appear, even at that time, to have made conſiderable progreſs. The knowledge, which Paracelſus carried home to his native country, was not generally known in Europe. This author applied it with ſucceſs in the cure of ſome deſperate diſeaſes; and acquired uncommon fame from his new and unheard-of remedies. He was an empiric in the theory, no leſs than in the practice of the art; and I may add, that his attempts to overturn the doctrines of the ancients, give an indication of more effrontery than genius or knowledge. The wonderful cures of obſtinate diſeaſes, which he was ſaid to perform; and ſtill more, perhaps, the myſteriouſneſs of his language, which caught the notice of the vulgar, who often imagine that knowledge is concealed under terms, which they do not underſtand, brought followers to his ſtandard. Thec-

ries of the proximate caufe of fevers, were fabricated
without difficulty, by the help of thofe principles,
which Paracelfus had introduced to the acquaintance
of the world; yet it does not appear, that any theory
arofe, during this period, which had probability, or
even ingenuity for its fupport. The period indeed,
during which chemical reafonings fo univerfally prevailed, may be ftyled juftly enough a period of medical romance: and I fhould confider it as a trefpafs
on the patience and good fenfe of the reader, to fpend
time in refuting the abfurd and incongruous doctrines
of fulphur, nitre or mercury; acid and alkali, or the
various modes of fermentations, which for a time
filled the writings of phyficians. The mechanical
mode of reafoning, which fucceeded, or rather which
became incorporated with the doctrines of the chemifts, feemed at firft to promife greater advantages:
but though theories of fevers were formed by many
eminent men, both of the laft and prefent century,
on the principles of the mechanical or chemico-me-
chanical philofophy; yet there are not any of them,
which feem to have afforded a fatisfactory explanation of the fubject.—The fo-much celebrated doctrine
of lentor and vifcidity was affumed without evidence
of its exiftence, and perfifted in, without being fufficient to account for the phenomena of the difeafe.

Thofe conjectures concerning the proximate caufe
of fevers, which I have mentioned hitherto, can feldom be faid to extend farther than to a particular
ftate of the humours, or circulating mafs of fluids,
which, according to the prevailing philofophy of different ages, have been fuppofed to be changed from
their natural and healthy ftate, by chemical or mechanical derangement. I obferved before, that it
might appear, from an accidental expreffion in the
writings of Hippocrates, that this author was not
altogether ignorant of the influence or effects of a
nervous power, or fentient principle. The methodic

H 3

fect likewife, may feem to have comprehended in the idea which they have given of difeafes, that there is fome derangement of the fibrous fyftem; or perhaps that a fpafmodic ftricture of capillaries is actually the immediate caufe of fever; whilft Galen every where celebrates the powers of nature or vires naturæ medicatrices, which bear in his opinion, a very active part in the cure of febrile difeafes. To thofe vague ideas of the ancients, we may add the more modern and explicit doctrine of Van Helmont, who was a man of genius, learning and obfervation. Van Helmont adopted the fentient principle of Hippocrates; but he alfo applied it in a bolder light than had been done by its original author, and employed its affertions more particularly towards the explanation of the caufe and phenomena of fevers. The enthufiafm of this writer difgufts the philofophic fpirit of the prefent age, and we muft acknowledge, that his ideas are often unphilofophical and abfurd; yet we muft likewife do him the juftice to add, that the principle of his doctrine in fome degree is well founded, and that his views, in many refpects, are important in practice. I muft further obferve, that the efforts of nature, fo celebrated by Campanella and Sydenham, and even, perhaps, the αυτεκρατεια, of Stahl and his followers, can only be confidered as modifications of the furious Archeus.—But though the authors I have mentioned, feem evidently to have poffeffed fome vague idea of the powers or influence of the nervous fyftem; yet there are not any of them, who have attempted to explain its operations by a philofophical and confiftent mode of reafoning. The period of this improvement is not very remote.

As foon as the circulation of the blood was known and fully eftablifhed, the heart loft fome part of its former importance; whilft the brain and nerves, which for many ages had been little regarded, rofe into primary and effential confequence. But though the brain and nerves were difcovered, foon after this period, to be the inftruments of life and motion; yet the laws of this part of the fyftem were at firft only imperfectly underftood; and the attempts to explain its operations were, for a while, whimfical and abfurd. Willis deferves fome credit, as being one of the firft who brought the general importance of the nervous fyftem into view: but Borelli, an Italian mathematician, actually appears to be the firft who ventured to afcribe the proximate caufe of fever, to a particular derangement of this part of the frame. The immediate caufe of fever, in this author's opinion, depends on fome unufual acrimony of the nervous fluid; but it is only neceffary to obferve with regard to this doctrine, that a fuppofition of acrimonious fluids, where a fluid cannot be proved to exift, is fo perfectly vifionary, as only to deferve to be mentioned, from its being the firft attempt to bring this part into view, in accounting for febrile difeafe. This hypothefis, however, though obvioufly ill founded, enjoyed its day of fame. It was foon followed by another conjecture, more ingenious indeed, but which was not fo generally attended to, as the preceding. Dr. Cole of Worcefter, towards the end of laft century, fuggefted an idea, that the proximate caufe of intermitting fevers depends on a laxity or debility of the brain and origin of the nerves. The fuppofition is not fo improbable; but the fuperftructure, which the author has raifed, is abfurd, and unfupported either by fact or probability. Yet, if ceptwe ~~except~~ Mundy, an author who offered a conjecture of a fimilar kind, in a work entitled Βιοχρησολογια, Borelli and Cole are the only writers prior to the time of Hoffman, who confidered the nervous fyftem, as directly affording a feat for the proximate caufe of fevers. Hoffman, whom I have juft mentioned, was a celebrated profeffor at Halle in Saxony. He flourifhed in the earlier part of the prefent century, publifhed many volumes, and certainly poffeffes the merit of having enlarged our views on the fubject of fevers. His theory of the proximate caufe is not only more ingenicus, but certainly has more appearance of truth, than any other, which had been offered to the public at the -time it appeared. The ~~caufe~~ of fever confifts, in his opinion, in a fpafmodic affection of the nervous fyftem. It is a truth which few people will attempt to deny, that a fpafmodic ftricture of the furface of the body generally takes place in ordinary cafes of fever ; yet we muft perhaps alfo acknowledge with Dr. Cullen, that a fpafmodic ftricture is not certainly and uniformly the firft effential part of a febrile difeafe. Some other thing is frequently obferved to precede the fpafm, which, in the opinion of the laft mentioned celebrated profeffor, has a right to be confidered as a proximate and effential caufe. But as the theory of the proximate caufe, affigned by Dr. Cullen, is not only more plaufible and complete than any preceding one; but ftill increafing in popularity and fame, it will not be fuperfluous, if we ftop to examine it with more attention. The remote caufes of fever, according to this author, are fedative powers, applied to the nervous fyftem, which diminifhing the energy of the brain, thereby produce a debility in the whole of the functions, and particularly in the action of the extreme veffels. Such, however, is at the fame time the nature of the animal economy, that this debility proves an indirect ftimulus to the fanguiferous fyftem ; whence by the intervention of the cold ftage and fpafm connected with it, the action of the heart and larger arteries is increafed, and continues to be fo, till it has had the effect of reftoring the energy of the brain, of exciting this energy to the extreme veffels, of reftoring therefore their action; and thereby fpecially overcoming the fpafm affecting them : upon the removing of which, the excretion of fweat, and other marks of relaxation of excretories take place. This theory of fever holds out an appearance of great fimplicity, and of perfect connexion. I wifh we could fay that it had an equal claim to truth: but I am afraid it will be found, on a careful examination, to be no more in reality than an ingenious hypothefis, the leading principles of which can fcarcely be proved even to exift. I do not pretend to enter deeply into the difcuffion of the fubject; yet I cannot avoid reprefenting, in a few words, fome circumftances of difficulty in this author's theory, which are not eafily reconcileable, either with reafon or obfervation. It might be doubted, in the firft place, if the remote caufes of fever are actually of a fimple fedative nature; but at prefent I fhall admit that the firft principle, which is affumed by the profeffor, is in reality a fact, and proceed to enquire, if the reft of the doctrine is capable of being defended, even on this foundation. It conftitutes the fum of Dr. Cullen's theory, as was mentioned before, that the remote caufes of fever occafion debility, or diminifhed energy of the brain and nervous fyftem; that this debility neceffarily gives rife to fpafm, and increafed action of the heart and arteries; which continuing for a certain length of time, finally removes the difeafe. Thus the different ftages of fever appear to follow each other as caufe and effect; and debility in the firft inftance, is fuppofed neceffarily to give rife to reaction. Such a fuppofition is not very obvious to reafon, and has not much fupport from the analogy of facts. It would be eafy to mention examples, .where the application of debilitating caufes as is not obferved to be followed by obvious reaction of the fyftem; but at prefent I fhall content myfelf with the familiar one of the application of cold. It is perfectly well known, that cold, when conftantly and uniformly applied to the body, even goes fo far as abfolutely to extinguifh the powers of life, in a part, or in the whole, without our being able to perceive any efforts on the part of nature to stop the progress of this destructive tendency. From this we may fairly infer, that common debilitating causes, at least while they continue to be applied in the same constant and uniform manner, do not necessarily excite the reaction of the system: but I will even go farther, and venture to affirm, that spasm and reaction do not necessarily follow very great degrees of debility, which appear to proceed from the presence of a febrile cause. During the time I remained in America, I had frequent opportunities of witnessing the truth of this assertion. In the southern provinces of that country, particularly in the summer and autumnal months, the intermitting fever was generally epidemic in a high degree; but its general cause, which was then so abounding in the atmosphere, often injured the actions of life, without producing a regular train of operation; that is, one part of the disease appeared without that mode of action, which is supposed, by our author, to be its necessary effect. Thus, I have seen the most extreme degrees of debility and languour in all the functions, continue even for eight or ten days, without our being able to discover the smallest marks of spasm, or obvious reaction. This inactivity and languour sometimes vanished suddenly; and the body resumed its ordinary health and vigour, frequently without an evident cause. On the next day, however, or perhaps the day following it, the patient was surprized with a regular paroxysm of fever. From this it appears very plainly, that if the immediate cause of fever actually consists in debility, this debility necessarily undergoes a peculiar, but hitherto undefined species of modification, before it can be considered as the cause of the subsequent parts of the disease; a concession which leaves us perfectly in our former state of uncertainty and ignorance.

As it may be concluded from the facts, which I

have mentioned, that fpafm and reaction are not the necelfary confequences of the application of debilitating caufes, either common or febrile; fo if we purfue our author's train of reafoning farther, we fhall not find his inductions to be very confiftent, or very convincing. If we are difpofed to grant, that the remote caufes of fever actually dimimifh the energy of the brain, it is not an obvious inference, that the circumftances of this diminifhed energy have the certain effect of exciting the reaction of the fyftem. It appears, in fhort, like afcribing reft and motion to the fame power. But to fmooth the prominent features of this apparent inconfiftency, the ingenious author has thought fit to affume a principle, the exiftence of which is very ambiguous in its enlarged fenfe, and very infufficient in its limited one. Dr. Cullen does not admit of the Italian principle of αυτοισατεια; he however afcribes effects to the vis naturæ medicatrix, which are not capable of being explained mechanically. I mentioned before, that no efforts of nature are perceived to arife, under the uniform and conftant application of a debilitating caufe; but though this is true, I muft likewife obferve, that when thefe debilitating powers, from any caufe whatever, actually ceafe to act, abate materially in the intenfity of their action, or fuffer change in its mode, before the vital principle is irrecoverably deftroyed, nature, which perfifts in continuing life, and even ftruggles in attempting to maintain it, may then be faid to raife efforts, which have a tendency to reftore the body to its ordinary health. This power, which to a certain degree, refifts caufes of a deftructive tendency and which endeavours to reftore to their original ftate the derangements of the fyftem which have actually taken place, is only a limited degree of the vis naturæ medicatrix. It is in fhort, no more than an effort to continue the action of living; yet it is all, which we fhall, at any time, perhaps, be able to perceive.

I have thus mentioned briefly some objections, to this celebrated theory of Dr. Cullen. The ingenuity of the author is acknowledged to be great; the pains and labour, which he has beftowed in completing his favourite doctrine appear likewife to be confiderable; yet I cannot help remarking, that its defects are ftill fo obvious, that we are unavoidably obliged to be fatisfied with one of thefe conclufions: viz. either that the debility, which is fuppofed to be the caufe of fever, is of a peculiar but unknown kind; that it ceafes to act, or changes its mode of action from an accidental caufe, or from fomething in its own nature; or that a reaction arifes in the fyftem, from a principle of confcioufnefs of the deftructive tendency of this debilitating power. It is not confonant with the common laws of the animal economy; that reaction fhould arife in that part of the fyftem, where the debilitating influence has been primarily and principally exerted. Suppofitions of this nature, are only fubterfuges, and no more in reality than myfterious ways of acknowledging ignorance.

The opinions I have enumerated above are the principal ones, which have been advanced by medical writers, on the fubject of the proximate caufe of fevers. Though numerous, they are all reducible to two general claffes; viz. either to conjectures, which are totally without foundation; or to circumftances, which are in fact only fymptoms or parts of the difeafe, fome of which are more, others lefs effential. The proximate caufe of fever, is a certain peculiar ftate of the body, on which the difeafe, or the fubfequent parts of the difeafe, neceffarily depend. It is, in fhort, the firft effential action of the febrile caufe; but this action is fo intricate and difficult to be difcovered, that phyficians have fought for it in vain for more than two thoufand years. The ancients were fatisfied with the idea of preternatural heat, excited in the heart, and communicated, by means of the blood and spirits, to the rest of the body. Hoffman, making a bolder step, introduces a spasmodic affection of the moving fibres; and Dr. Cullen, going still farther, lays the principal stress upon languor and debility, or weakened action of the nervous energy. Increased heat, spasmodic stricture and marks of debility are generally present, in various degrees, in the different stages of fever; but debility for the most part precedes the others; and on this account, if equally essential, has a preferable right to be considered as the first part of the disease. There is still reason to doubt if it actually is the first. I have myself attended carefully to the manner in which intermitting fevers approach. The first thing which I observed in others, or what is still more to be depended upon, the first thing I felt in myself, was usually a disagreeable, but a peculiar affection of the stomach. The precise nature of this affection I am unable to define in words; but I knew it so well by experience, that I always considered it as a warning, (and it was sometimes the only warning, which I had,) of the approach of the paroxysm. It was often accompanied by flatulence, and it sometimes preceded the first feelings of the languor and debility, nearly the space of an hour. The observation of this fact has occured to me frequently; and I cannot avoid concluding, that it gives room to believe, that the debility, which is so commonly the fore-runner of fever, instead of being the first and principal mode of action of the febrile cause, is only a part of that action, —perhaps not the most essential. As Hippocrates appears at a very early period to have been perfectly sensible, that something besides a simple increase of heat was necessary to constitute a fever; so we may now perhaps conclude, with equal reason, that debility has some other circumstances combined with it, which we have not been able to discern very clearly.

I

Having faid that there are not any of the numerous theories, which have as yet been offered to the public on this important fubject, in any degree fatisfactory, it might be expected, perhaps, that I had fomething of my own to bring forward, which might be more perfect, at leaft in my own opinion : but I willingly acknowledge, that I have no fuch pretenfions. After fourteen or fifteen years of careful obfervation, and daily reflexion on the phenomena of fevers, I am obliged to confefs, that my opinion ftill remains to be formed. The proximate caufe of this difeafe, is a fubject of a dark nature. It is fuch, perhaps, as our limited capacities will never develope. But though we defpair of ever attaining clear ideas of its fpecific nature, there are ftill fome ufeful circum-ftances connected with it, which we comprehend with clearnefs. We know, that the more general remote caufes of fevers, are certain invifible exhala-tions, fometimes more evidently arifing from marfhy grounds, fometimes more obfcurely diffufed in the air, and fometimes obvioufly proceeding from the bodies of our fellow creatures. We know, likewife, that thefe caufes which are unfriendly to the human conftitution, are varioufly modified and combined, and of various degrees of force or in various ftates of concentration ; but we proceed no farther with cer-tainty. We are not able to afcertain the nature of thefe effluvia, and it is only by conjecture that we . trace them in the channels by which they enter the body. The changes which they operate on the folids, fluids, or nervous fyftem, before their action becomes obvious, are totally unknown to us. We know, though the body lie expofed to exhalation, even in a concentrated ftate, that an appearance of difeafe is not, generally, the inftantaneous confequence. A fpace of time intervenes, various indeed according to circumftances, but always fuch as gives room to believe, that the caufe requires, and actually under-

goes a modification, before it is capable of producing
a fever, or the paroxyfin of a fever. The circumftances connected with the approach of fevers, particularly intermitting fevers, afford an illuftration of
my meaning. The caufe of the difeafe, fo far from
producing the fever immediately when applied to the
body, often lurks for a confiderable time in the conftitution, without perceptibly injuring the ordinary
actions of life. Sometimes it gives rife to affections,
which are apparently very different from their real
nature. Thus a perfon often languifhes for days,
weeks, or even longer. The indifpofition fuddenly
and unexpectedly vanifhes : and the apparent recovery of health is foon followed by a paroxyfm of regular fever. In other cafes again, the attack of the
difeafe is fudden; and its formation from the beginning diftinct. This fact affords a prefumption, that,
in confequence of a particular modification, which is
only accomplifhed in a certain fpace of time, but
the nature of which we do not in the leaft comprehend, an aptitude is regenerated between the remote
caufe of the difeafe, and the relative ftate of the body.
When the ftate of the body, and the remote caufe approach to, or arrive at a ftate of mutual correfpondence, the difeafe is produced. When this ftate, which
I call an aptitude, is changed or deftroyed, the difeafe
vanifhes, or fuffers a change of form. This is a fact,
which cannot be difputed; and it feems to be the
extent of our knowledge, on the important fubject of
the proximate caufe of fevers.

# CHAP. VI.

IT will not be improper to remark, before I begin to defcribe the hiftory of this fever, that though the endemic which prevailed at Savanna la Mar, notwithftanding much variety of forms and fymptoms, is confidered as only one and the fame difeafe, yet it may alfo be obferved, that the variety of thefe appearances is fometimes fo great and remarkable as to occafion confiderable perplexity to the practitioner. The fymptoms and form of this endemic appeared, on a fuperficial view to be conftantly varying ; yet by attending more clofely to the courfe, progrefs, and changes of the difeafe, thefe apparent irregularities vanifhed gradually, the varieties being in fact only accidental, and often depending on very trivial caufes. The analyfis of the different cafes of fever, which came under my care, during the time that I lived in Jamaica, furnifhed me with this information. I formerly mentioned the manner in which my obfervations were conducted : I have only now to add, that I truft the method, which I adopted, has enabled me to give a more accurate hiftory of the fever of Jamaica, and to explain more fatisfactorily than has been done hitherto the various fources of the many irregularities which are obferved to occur. I am perfectly fenfible that my experience has been too limited, to give me hopes of rendering the prefent work complete; yet I would flatter myfelf, that it will not be altogether ufelefs : I totally difclaim theoretical opinions, and content myfelf with a plain narrative of facts; neither do I afpire to any higher praife, than care in obferving

the appearances of the difeafe, and truth in relating
the appearances which I have feen.

Before proceeding to give a particular hiftory of
the different varieties of the endemic fever of Jamaica,
it will not be improper to mention the more general
fymptoms, which diftinguifh the difeafe, and to trace
an outline of the courfe, which it has been commonly
obferved to purfue: and I may remark in the firft
place, that though debility is ufually confidered as the
firft fign of an approaching fever; yet, if we attend
minutely to all the circumftances of invafion, it will
not generally be difficult to perceive, that a difagreeable, though undefcribable affection of the ftomach,
takes place previous to the fmalleft perceptions of
languor or debility, which are commonly only immediate fore-runners of coldnefs and fhivering. This
coldnefs, which was obferved to be various in duration, as well as in degree of force in the fevers of
this country, was fucceeded by flufhings of heat alternating with the cold, and increafing gradually till
the heat at laft prevailed. The hot fit, which was
likewife of various duration and of various force, had
many new fymptoms joined with it, the principal
of which were fuch as fhewed an increafed circulation, or an irregular determination of the blood to
the different parts of the body. This hot fit, and
the difturbances connected with it, according to circumftances, continued for a longer or fhorter fpace
of time; at laft fweat breaking out on the head and
breaft, extended itfelf gradually to the extremities,
and accomplifhed after a certain continuance, either
a total remiffion of the fever, or a confiderable abatement of the violence of the fymptoms. It is almoft
unneceffary to mention, that this remiffion or abatement of fymptoms was of longer or fhorter duration,
and more or lefs complete in fevers of different forms.
An aggravation of fymptoms fucceeded to the re-

I 3

miffion; but it was ufually obferved to begin without preceding coldnefs, and frequently without marks of preceding languor or debility. The hot fit now ran high, and all the fymptoms were frequently more violent than they had been obferved to be in the firft paroxyfm. Sweat at laft made its appearance, followed in moft cafes by a remiffion, lefs perfect indeed, than the preceding one, but ftill diftinct enough to be clearly traced. In this manner things went on for a longer or fhorter fpace of time, the paroxyfms ufually increafing in violence, and the remiffions becoming fometimes more, though in general lefs perfect, as the difeafe advanced in its progrefs. I may further obferve, that there was occafionally a change of the type, fometimes a change of the nature of the fymptoms in the courfe of the illnefs; and that, where either of thefe were the cafe, the difeafe was ufually of longer continuance; at the fame time, that the order of the critical days was difturbed in confequence of thefe changes.

The refemblances, which I have mentioned above, were found in all the different fpecies of the remitting fever of Jamaica; but from caufes, which were not always perceived, and which fometimes appeared to be very accidental, the difeafe was diftinguifhed in a part, or in the whole of its courfe, by the prevalence of a train of fymptoms of fuch a particular nature, as gave occafion to the diftinctions of inflammatory, nervous, malignant, putrid or bilious; the feparate hiftories of which I fhall now relate more circumftantially.

## SECTION I.

### OF FEVER DISTINGUISHED BY SYMPTOMS OF IN-FLAMMATORY DIATHESIS.

THE variety of fever, which I shall describe first, is that, where the inflammatory diathesis prevailed in different degrees. Where this diathesis was moderate, the disease was usually of the least complicated form, as well as of the least dangerous nature, of any of the fevers of Jamaica. The paroxysms were generally regular, and complete in all their parts, and terminated, for the most part, by a copious sweat, in a perfect remission : the pulse was full, strong, and regular ; without uncommon hardness or tension ; whilst the heat of the skin, though sometimes great in degree, was generally free from that burning pungency, so common in some other species of fever. It was less removed, in short, from a simple increase of the natural warmth. I may further remark, where this moderate degree of inflammatory diathesis characterized the genius of the disease, that the danger was seldom great; and that the termination or crisis was generally regular and final. But though this degree of the inflammatory diathesis was frequently observed to be a sign of safety, and of regular crisis : yet it also often happened, where the diathesis prevailed in excess, that the symptoms of excitement ran unusually high, and that a serious danger threatened life. The pulse, in such cases, was not only frequent during the paroxysm, but it was likewise, quick, hard, and vibrating ; the heat was often intense ; the internal functions and the various secretions, were considerably disordered ; at the same time, that a very obstinate spasmodic stricture prevailed on the surface of the body. The remission which followed, for the most part, was obscure; the pulfe frequently retaining a preternatural quicknefs and hardnefs; whilft there was generally a confiderable degree of febrile heat on the fkin.

The fymptoms, which I have juft now mentioned indicate different degrees of the real inflammatory diathefis ; but befides actual fymptoms of real inflammatory diathefis, there were likewife found fevers, with the appearances of a fimilar difpofition, though the real genius of the difeafe was in reality of a different nature. It is of importance in practice to diftinguifh thofe ambiguous appearances ; but it is not always eafy to do it with certainty. We may remark, however, that the apparent inflammatory diathefis was ufually accompanied with marks of great irritability, and fometimes with marks of violent excitement during the paroxyfms ; while languor and great depreflion of fpirit were frequently perceived to attend the remiffions. The pulfe, which at one time was hard, irregular, and quick, at another was frequent and low, and funk under a fmall degree of preflure. The heat of the body was not always great, yet it was pungent,—and left a difagreeable fenfation on the hand : the fecretions were often irregular ; the countenance was confufed, clouded and overcaft, the eye was fad, and fometimes appeared as if it were inflamed ; the feelings were unpleafant to the patient himfelf : there was great irritability of temper; and the ftate of the fkin impreffed us with the idea, that there was a ftrong fpafmodic ftricture prevailing on the furface of the body.—The above are the principal circumftances, which were ufually prefent in the different ftates of inflammatory fevers; yet thefe circumftances were fometimes fo varioufly complicated and combined with others, that the accurate difcrimination of them muft be left, in moft cafes, to the obfervation of the individual himfelf.

Thofe different ftates and degrees of the inflammatory diathefis, which I have defcribed above, were fometimes general throughout the whole of the body, not affecting one member more remarkably than another; fometimes they were partial or feemed to be connected with a principal affection of a particular part : and where this was the cafe, the local affection, and the general diathefis of the fyftem, ufually had a mutual correfpondence with each other. Thus, where the inflammation affected the fubftance of the liver or lungs, the general inflammatory diathefis was ufually in a moderate degree ; while the higheft excefs of general vafcular excitement often accompanied inflammations of the membranes of thofe organs. But though inflammation of membranes was often accompanied with a high degree of general inflammatory diathefis; yet there were likewife fome. kinds of thofe local inflammations which communicated only a low, or an ambiguous degree of their diathefis to the general fyftem : fuch are fome of thofe inflammations, which occafionally affect the furface of the alimentary canal, and which appear, in general, to be of the eryfipelatic kind.

## SECTION II.

### OF FEVER WITH SYMPTOMS OF NERVOUS AFFECTION.

THE endemic fever of Jamaica, was not oftener diftinguifhed by fymptoms of general inflammatory diathefis, than by circumftances of nervous affection. The beginning of this form of the difeafe, was often characterized by a high degree of that difagreeable affection of the ftomach, as alfo by much of that languor and debility, which are commonly fore-runners of fevers in general. To thefe fymptoms fucceeded a flight degree of chillinefs, followed by a hot fit, which often continued long, but feldom ran high. The pulfe was fmall, frequent, and eafily compreffed. It varied with change of pofture;— and fometimes was fo much affected when the patient was raifed upright, as totally to difappear; the heat of the body was feldom great; the fecretions and exertions were generally irregular, and the internal functions were much difordered. The mind was ufually affected, affected however in various degrees, and in various ways. Sometimes there was a lively delirium, fometimes the delirium was low and defponding; and, as the one or other of thefe was the cafe, the appearance of the eye and countenance was chearful or fad. The tongue was fometimes moift, fometimes dry, but feldom very foul; thirft was irregular, naufea was frequent, and the ftate of the ftomach was generally very irritable. There was likewife, in moft cafes, deep and heavy fighing, and, unlefs in times of preternatural excitement, a very urcommon degree of defpondency. The above were the principal fymptoms of the nervous fever of Jamaica. The paroxyfms in this difeafe feldom exceeded twelve hours in duration; while the termination or abatement, was ufually diftinguifhed by fweating, though feldom by fuch fweatings as extended completely to every part of the body. The remiffions were not by any means perfect: the head-ach, and other difagreeable feelings ufually abated; but figns of languor ftill continued, and marks of fpafmodic ftricture for the moft part remained on the furface of the fkin. I may further obferve, that as the paroxyfms generally increafed in violence, in the progrefs of the fever; fo it was very feldom that the remiffions put on an appearance of greater diftinctnefs, as the difeafe approached to its termination.

Such is the general hiftory and the progrefs of the difeafe, which might be diftinguifhed by the name of the nervous fever of Jamaica: but befides thofe cir-

cumftances, which I have mentioned above, others
were fometimes found attending it, which, though
lefs regular and conftant, deferve ftill to be taken notice of. Thus the firft ftage of the paroxyfm, inftead
of the more ufual appearances, was occafionally diftinguifhed by fits, which appeared to be of the epi-
leptic kind. Thefe fits in fome cafes were fucceeded
by a lively delirium, in others by ftupor or infenfibility. The delirium, which was a common fymptom
of this difeafe, ran high in feveral inflances; though
it more generally amounted only to an abfence of
thought, or difficulty of recollection. It is a circumftance of fome curiofity likewife, that inftead of a
paroxyfm, confifting of different parts in a certain
order of fucceffion, there was fometimes a total ftupor
and infenfibility, which continued for a determinate
fpace of time, without even being fucceeded by obvious marks of fever: whilft the time of the paroxyfm,
in other cafes, was diftinguifhed by fuch a degree of
tremor and mobility, as nearly approached to the
difeafe known by the name of St. Vitus's dance.
And further, befides thefe ftrange and irregular appearances, fpafms and excruciating pains in different
parts of the body, in many inftances, were the leading, indeed almoft the only fymptoms of the difeafe.
 It is not only curious, but it is indifpenfably neceffary in the conduct of our practice, to obferve with
attention the various modes of action of the caufe of
fevers, and to eftimate with precifion the various
combinations. The caufe of fevers, in exerting its
principal action on the nervous fyftem, fometimes
produces excitement, fometimes occafions depreffion;
effects oppofite to each other in their nature. Excitement and depreffion are two general and oppofite
modes of action; yet befides thefe we often obferve
others, which do not belong wholly to the one or the
other, but which feem to be compounded of both, in
a manner we do not very well comprehend. This

caufe of fever likewife, which acts in directly oppofite ways, appears alfo to exert its action more powerfully at different times on one part of the fyftem
than on another; that is, it acts fometimes more immediately on the brain, or reafoning faculty, fometimes more directly on the nerves, or moving powers
of the body. It may even be obferved further, that
all thefe modes of action, which are preferved diftinct
at one time, are combined in various degrees at another. Thus, where the caufe of fever acts by producing excitement, lively delirium in various degrees
is the confequence; while languor, ftupor, and infenfibility naturally follow the oppofite mode of action. Low delirium, tremors, ftartings, &c. are probably owing to a compound effect. Both modes of
action fucceed each other rapidly; or perhaps both
modes are actually prefent at the fame time, though
probably in different degrees, in the different portions of the brain. This fact at leaft is certain, that
obvious depreffion is often combined with figns of
great irritability. It is a remark likewife of confiderable importance, that the natural functions are lefs
difordered, where the caufe of the difeafe acts upon
the nervous fyftem internally, or principally difturbs
the intellectual powers, than where this action is obvioufly external: the pulfe is then more regular, though
often obfcure; the difpofition to faint is not fo great;
mufcular mobility is lefs remarkable, and local pains
are felt lefs acutely. On the contrary, where this
caufe acts externally, or chiefly affects the moving
powers, the difpofition to faint in changing pofture
is more remarkable; tremors, ftartings, &c. are more
common; appearances, in fhort, are more fluctuating and often more alarming.

It is a matter not lefs ufeful than curious to diftinguifh the different fpecies of delirium in fevers, to
trace the different combinations, and to mark the apparently trivial caufes, which excite, or which fome-

times remove thofe derangements of the reafoning faculty. It is a remark, which has been often made,
that while one delirious perfon in fever appears only
to be in better fpirits than ufual, another, or perhaps
the fame perfon in another paroxyfm of the fame difeafe, is outrageous or perfectly furious. A third is
low and languid, abfent and inattentive, or, with a
fixed look of vacancy, does not feem to be otherwife
deranged, than by requiring greater time to recollect
himfelf. To which we may add, that there are fome,
who talk coolly on things in general ; but who cannot
bear mention of fome particular fubjects.

---

# SECTION III.

### OF FEVER IN WHICH ARE DISCOVERED SIGNS
### OF MALIGNITY.

THE fever of Jamaica, as diftinguifhed by figns
of inflammatory diathefis, or by circumftances
of nervous affection, prevailed principally at Savanna la Mar ; yet befides the above forms of the difeafe, there fometimes likewife occurred others, which
fhewed marks of peculiar malignity. It is difficult
to define precifely in words the character of the difeafe, which I now attempt to difcribe ; its difcriminating marks, not confifting fo much in one or two
fymptoms, as in a certain affemblage of circumftances
refiding chiefly in the ftate of the eye and countenance of the patient, and conveyed with difficulty
in verbal defcription. I may remark, in the firft
place, that there was feldom any thing very particular in the manner of invafion of this fpecies of difeafe. The cold fit was rarely violent in degree, though
it was often of long continuance: neither did the hot

K

fit ufually run high, in the common acceptation of the
word, though it was fometimes attended with circumftances peculiarly difagreeable. The pulfe varied remarkably. It was obfcure, or fcarcely to be felt in
fome ; in others it was ftrong, though unequally fo;
the artery, in many inftances, being hard and contracted, with a peculiar vibration in the ftroke. After thefe fymptoms and others, which are ufual in this
ftage of fever, had continued for a longer or fhorter
time, fweat began to make its appearance on the head
and breaft, which extending itfelf gradually to every
part of the body, was at laft followed by a remiffion,
tolerably perfect for the moft part, though there ftill
remained fome ftrange and unpleafant fenfations. It
does not appear that there is any thing very uncommon in the fymptoms, which I have hitherto taken
notice of: thofe which follow are more characteriftic.
The ftate of the eye and countenance, afford the fureft
figns of the malignity of the difeafe ; but there is difficulty in difcriminating thofe appearances. The face
is not unufually flufhed in fevers; but, in the prefent
cafe, the countenance exhibits fomething elfe befides
an appearance of fimple flufhing. It is likewife grim,
dark and overcaft, with fuch marks of confufion and
diftrefs, as if the patient were agitated by fome re-
fentful paffion. The eye is fad and defponding; and
the whole appearance, in fhort, indicates fuch a ftate
of mind, as we fhould be difpofed to ftyle malignant.
It is in fuch a ftate of the countenance as I have defcribed, that the character of this fpecies of fever
chiefly refides ; yet befides this, fome other circumftances frequently attend the difeafe, which are lefs
ufual in ordinary fevers. The paroxyfm for inftance
returned, for the moft part, much fooner than the
regular period, always with greater violence, and
fometimes with new and alarming fymptoms. It declined in twelve or fourteen hours; but the remiffion
was lefs perfect than the preceding one; the next re-

turn of fever, which was likewife much earlier than the ftated hour, was often ufhered in by convulfions, and the time of it occupied by ftupor or coma. The tongue was likewife irregularly moift or dry. If dry, it was generally covered with a black fcurf; if moift, with a thin glutinous coat, through which the red furface fhining obfcurely, prefented an appearance of a leaden colour. In this cafe the mouth abounded with a ropy faliva. But befides the above fymptoms, there were alfo violent twitchings in the ftomach and bowels, fudden fqueamifhnefs, faintnefs, anxiety, reft-lcffnefs, frightful dreams, diftreffing apprehenfions, and frequently after the fecond paroxyfm, a particular crouded eruption (not unlike iron-burnt blifters,) on the upper lip, which for the moft part fpread towards the nofe. The type of this fever, it may be further remarked, was ufually of the fingle tertian kind, generally anticipating by long anticipations. In moft inftances this malignant difpofition was difcoverable at the very beginning ; yet in others, no fymptoms of a doubtful nature made their appearance till after the third revolution.

# SECTION IV.

## OF FEVERS IN WHICH ARE OBSERVED SYMP-TOMS OF A PUTRESCENT TENDENCY.

WE meet with the term putrid fever, or fever with putrefcent tendency, in the writings of almoft every author who has treated of the difeafes of hot climates: but though this expreffion is fo much the common language of practitioners, I cannot help obferving, that a remitting fever, with fymptoms of a fpecific putrefaction, did not once occur to my

obfervation in the ifland of Jamaica, during the time
that I lived in that country. I muft however add,
,that though a remitting fever fpecifically putrid is actually a rare difeafe; yet I do not attempt to deny,
that a putrefcent tendency is frequently prefent in the
primæ viæ, in a very confiderable degree ; and that
marks of it are fometimes difcoverable, even in the
general fyftem, at a late period of the illnefs, when
the vigour of life has abated, and the powers of cir-
·culation have begun to fail. This however is fo accidental and uneffential, that it is only in compliance
with the general language of medical people, that I
think it neceffary to defcribe a difeafe, where thefe
fymptoms are obferved to prevail. The tendency to
putrefaction, which was obferved in the fever of Jamaica, fometimes begins in the primæ viæ; and from
the primæ viæ was communicated to the reft of the
fyftem ; fometimes it remained confined to the limits
of the inteftinal canal, throughout the whole duration of the diforder; in which cafe flatulence, ructus,
anxiety, naufea and thirft were the fymptoms which
were chiefly troublefome: the belly likewife was generally loofe, at the fame time that the ftools were
dark and fetid. But where this tendency was communicated from the primæ viæ to the reft of the body,
or othewife made its appearance in the general fyftem,
a form of difeafe arofe diftinguifhed by the following
fymptoms. If the tendency to putrefaction appeared
at an early period, the heat of the fkin made a more
difagreeable impreffion on the hand, than was ufual
in fome other fevers ; the fkin itfelf was likewife for
the moft part, dry and conftricted ; the thirft was irregular, fometimes intenfe, fometimes from local affection of the fauces, apparently little increafed.—
The appearance of the eye was often fad ; fometimes
it gliftened with unufual brilliancy; fometimes it feemed to be inflamed. The countenance was generally
flufhed, often particularly confufed, and of a grim

and clouded afpect. I have however frequently obferved, where fymptoms of putrefcency difcovered
themfelves at a late period of a fever, the preceding
courfe of which had been diftinguifhed by circumftances of nervous affection, that the bloom of the
complexion was uncommonly fine and delicate. To
the above fymptoms might be added, great irritability of temper, general uneafinefs of fenfation, and
diforder in all the functions of the body. When the
fever affumed this appearance, paroxyfms and remiffions were generally obfcure and irregular. The fever indeed often fubfided in a fmall degree; but the
future remiffions generally became lefs diftinct, as
the difeafe proceeded in its courfe. The tongue affumed different appearances, at different periods and
in different perfons. In fome it was moift, in others
parched and dry. It was not univerfally foul, at leaft
it frequently happened, that the edges were clear and
beautifully red in their colour. The lips likewife
were fometimes fmooth, and of a cherry-like appearance ; at the fame time that the gums were inflamed
and fpongy, as they ufually are in fcurvy : the pulfe
likewife was fmall for the moft part; but it was irregularly fo. I fay nothing of the difpofition to faint
in erect pofture, which though generally enumerated
among the figns of putrid fevers by authors, does not
in fact appear to conftitute a criterion of the difeafe.

K 3

# SECTION V.

## OF FEVERS ACCOMPANIED WITH AN INCREASED SECRETION OF BILE.

THOSE fpecies of fevers, which I have mentioned above, feem to affect the general fyftem, or every part of the body nearly alike; but befides thefe, we fometimes meet with others, which are diftinguifhed by local affections, or increafed determinations to particular parts in a degree fo remarkable, as to perfonate very exactly a peripneumony, a hepatitis, or inflammation of the bowels; the accompanying fever being at the fame time fo flight, as fcarcely to be confidered as a primary affection. As an accident fimilar to thefe local affections of the liver or lungs, we may reckon an increafed fecretion of bile. The caufe of fever, from circumftances which we do not always perceive, fometimes acts with particular violence on the biliary fyftem, in confequence of which the fecretion of bile being preternaturally increafed, a difeafe arifes, which without much impropriety may be called bilious. But though this irregular action of the morbid caufe, on the biliary fyftem, frequently gives rife to bilious appearances in the fevers of Jamaica; yet thefe appearances are in fact often owing to caufes more accidental, and more remote than even this. Naufea and vomiting are among the common fymptoms of fevers in every country; but they are particularly frequent in thofe of the Weft-Indies. It is well known that a continuance of naufea, or that a repetition of the action of vomiting, increafes the determination, not only to the ftomach, but likewife to the parts which are near to it. Hence the fecretion of bile is preternaturally increafed fecondarily by the ordinary effect of vomiting, and bilious appearances become a

neceffary confequence of this accidental fymptom of
the difeafe. In thofe two manners, viz. in confequence
of the irregular action of the morbid caufe on the
immediately biliary fyftem, or from a fecondary effect in confequence of its action on the ftomach, the
the bilious fever may, in fome refpects, be confidered
as a difeafe of nature; but befides this, it often originates from our own treatment, viz. from the repeated ufe of emetics, or of cathartics, which are violent in their operation. The accidental appearance
of bilious vomitings, in the fevers of hot climates, furnifhed medical authors with a pretence of forming a
new theory, and of directing the mode of practice to·
a particular view. Influenced by this appearance,
they affume it as a fact, that a vitiated quality, or a
redundant quantity of bile conftitutes the effential
caufe of the difeafe ; and on this foundation adopt
the plan of repeated evacuating, both upwards and
downwards ; a practice which evidently increafes the
fecretion of the bile. Hence, a difeafe, or the fymptom of a difeafe, arifes wholly from this mode of treatment ; and the removal or cure of it is afterwards
attempted by a perfeverance in the means, which originally gave rife to it :—of this I have feen numerous
examples.

I have now defcribed the remitting fever of Jamaica, as characterifed by fymptoms of a different
appearance. I may further remark, that where thefe
fymptoms were unmixed with each other, there was
little difficulty in the diftinction, and little embarraffment in planning or executing the indications of cure:
but it fometimes alfo happened, that the different fpecies, which I have defcribed feparately, was fo perplexed and complicated, that it appeared uncertain
to which kind the difeafe properly belonged; or to
which view the practice ought to be principally directed. Symptoms of putrefcency, for inftance, were
often combined with fymptoms of apparent inflam-

matory diathefis; as fevers with nervous affection,
or putrefcent tendency, were fometimes accompanied
with marks of peculiar malignity.   It happened often
likewife, that the nature of the difeafe fuffered a total
change after a certain duration ; or that a fever with
one train of fymptoms ceafed, whilft another with a
different train began.

It would be a matter of no fmall importance, were
we able to afcertain the various caufes, which influence the various appearances of the fame difeafe; but
this knowledge is not eafily attained :—much of it
indeed lies beyond the reach of our comprehenfion.
We may however remark, that the feafon of the year
ufually has fome effect on the diathefis of the fyftem,
and often on the type and form of the fever.   Thus,
in the dry feafon, though the remiffions are not always
more perfect, the type is commonly more fimple, and
the general diathefis is oftener inflammatory.   In the
rainy months, on the contrary, remiffions are more
perceivable, but the type is more complicated, and
the general diathefis of the fyftem has a ftronger tendency to putrefcency, often with a mixture of fymptoms of nervous affection, fometimes with fymptoms
of a malignant nature.   The ftomach, bowels, and
biliary fyftem likewife fuffer more in this feafon than
in the drier months of the year.   But befides this
difference, which arifes from feafon, we alfo find very
conftant effects from local fituation.   Thus in hilly
countries there is generally more of the inflammatory
diathefis, with more frequent determination to the
head and lungs, and lefs obvious remiffions, than in
flat and champaign countries, where the ftomach
and biliary fyftem fuffer in a more peculiar manner.

# C H A P. VII.

TO be able to perceive at a diftance, the approach of danger or returning health, is a knowledge highly fatisfactory and ufeful to the phyfician; but it is a knowledge which is not eafily attained : for to judge with certainty of the event of fevers, requires not only long and attentive obfervation, but a difcrimination of complicated and ambiguous appearances, which does not depend always upon attention alone. The fagacious Hippocrates is generally confidered as the firft, who laid the foundation of the fcience of prognoftic; and we certainly muft allow, that he has left us many important and valuable obfervations on the fubject; yet we may alfo add, that his decifions in many inftances, are precipitate. Hippocrates feems generally to have placed too great confidence in figns feparately confidered, and to have formed his conclufions too often on the authority of fingle facts. Thus he has fometimes confidered as fatal in themfelves thofe figns, which in reality are only dangerous. The abfolutely fatal figns in fevers are actually few in number. I am able to affirm, from my own experience, that people are fometimes reftored to health after many of the ufually reputed fore-runners of death are prefent. We have, in fact, as yet only an imperfect knowledge of prognoftic in fevers; but the field is ftill open, and careful obfervation, it is to be hoped, may enable us in time to fupply the defects. I dare not venture to affert, that I have advanced beyond others in this neceffary and difficult fcience; but I am difpofed to flatter myfelf, that the following attempt to appreciate the marks of danger or fafety in the fevers of Jamaica, may be found in fome degree ufeful. It contains the refult of my own obfervations in that country; and though I am perfectly confcious that the rules are often defective, yet I likewife know, that I have fuggefted fome hints which have not been commonly obferved, and which may help to direct thofe, who have not had much experience of their own.

Prognoftic is fuch, as applies to fevers in general, or more particularly to the different fpecies of the difeafe. The type or form, the general courfe and tenor of the diforders, and the *general* nature of the paroxyfms often afford ufeful information. From the type alone, we do not often obtain much that is to be depended upon. Long and diftinct intermiffions are commonly accounted figns of fafety; yet we frequently fee inftances of the fingle tertian proving fatal, while types of greater complication are often void of danger: in general, however, complicated types are fufpicious—and perhaps more commonly fatal than others. But though a knowledge of the type of the fever abftractedly confidered, does not commonly afford any material indication of danger or fafety, yet the time of the return of the paroxyfm is a fubject, from which more may be learned. An anticipation of an hour or two, is feldom much to be regarded; yet an anticipation of ten or twelve is always fufpicious. It either indicates a latent malignity, or a tendency in the difeafe to change to a continued form. The complication of another fever, or the doubling of type is by no means favourable; yet it is much lefs to be dreaded, than a long and an irregular anticipation. Anticipations have been generally confidered as figns of the increafing force of the fever; fo the type which poftpones, is ufually believed to indicate a difeafe, which is haftening to a favourable termination: the effect however is fometimes the contrary. I have myfelf feen fome inftances, where, in confequence perhaps of weaknefs and impaired fenfibility, the return of the fatal paroxyfm, though it probably had commenced fooner, was not clearly perceived till after the ufual hour of attack. But befides thofe indications of danger or fafety, which may be drawn from the nature of the fimple type, or from the hour of return of the paroxyfm, the ftate of the paroxyfms and remiffions deferves likewife to be attended to. It was generally obferved where the paroxyfms were regular, and affumed a completer form in the progrefs of the difeafe, that there was not generally much reafon to dread an unfavourable event. Hopes of fafety might likewife be entertained with ftill greater confidence, where the paroxyfms, though more violent in degree, became more regular and diftinct after the ufe of bark, wine and ftimulants. On the contrary, it was always an indication of danger, where they became longer or loft the diftinctnefs and regularity of their form, as the difeafe advanced in its progrefs. Changes from bad to good, in the courfe of the fever, alfo indicated more fafety as the oppofite changes indicated more danger, than if circumftances equally unfavourable had continued from the beginning.

In enumerating thofe particular figns or fymptoms, from which we are led to form a judgment of the event of the remitting fever of Jamaica, I fhall confider in the firft place the ftate of the pulfe. The pulfe is fo differently affected by the fame caufes in different people, and individually fubject to fo many peculiarities, that conclufions formed folely upon this bafis muft ever be fallacious. Hippocrates, who has treated very fully of the other figns of prognoftic, is totally filent on the fubject of the pulfe. He has mentioned the term, indeed, in feveral parts of his works; but it does not appear, that he had a perfect knowledge of the nature and indications of the pulfations of the arteries. The fubject was fomewhat better underftood before the time of Celfus: yet this author does not believe, that any information could be drawn from the ftate of the pulfe alone, which was in any great degree to be depended upon. Galen, who is generally diffufe on every fubject, has treated very fully of the nature of the pulfe. He has indeed multiplied diftinctions to an amazing extent, and fuggefted combinations of endlefs variety; yet notwithftanding this apparent minutenefs, there are ftill feveral important obfervations with refpect to it, which have efcaped him altogether. It is not many years ago, that Dr. Solano, a Spanifh phyfician who practifed at Antequiera, opened fome new and curious views concerning the pulfe, and its various indications.

The detail of facts with which this writer has furnifhed us, is really wonderful, and the candour with which he has related them, independent of the teftimony of feveral refpectable authorities, engages us to give him credit. I had not heard of Solano's difcoveries at the time I lived in Jamaica, and I do not find that I had ever taken notice of obfervations fimilar to thofe he has mentioned. I was able indeed, for the moft part, to foretel from the nature of the pulfe, even in the beginning of the difeafe, whether the fever would be of a continued or remitting form ; but I did not difcover any figns from it, which led me to form a judgment of the future mode of termination. I may add, that I met not with any inftances of crifis by hæmorrhage ; neither did I ever take notice of the rebounding pulfe. The intermitting pulfe occurred frequently, fometimes as a forerunner of of death, fometimes as an attendant of favourable crifis : but I cannot fay, that I obferved that it ever prefaged a future diarrhea. I fhall however pafs over the obfervations of others without further comment for the prefent, and content myfelf with relating thofe circumftances of pulfe connected with danger or fafety, as they occurred to my own obfervation in the remitting fever of the Weft Indies. I muft remark in the firft place, that independent of peculiarities of conftitution, a weak, a feeble and eafily compreffed pulfe was generally a bad one : a pulfe which was indiftinct and fmall, or fmall and hard, particularly at a late period of the difeafe, or together with delirium or clammy fweats, indicated for the moft part, the moft extreme degree of danger. That fpecies of pulfe moreover, where the ftroke was obfcure, or felt with difficulty, was fufpicious at all times; but it was particularly dangerous where accompanied with a wavering, a tremulous, a conftantly creeping or vermicular motion in the artery. I am not certain that my meaning will be clearly underftood; yet I believe that thofe who have once obferved this tremulous and creeping pulfe, will not eafily forget the danger which it indicates. It often attended a fever of a malignant kind, where the nervous influence appeared in fome degree, to be fufpended.—But to proceed : it is an obfervation fo well known as to render any mention of it almoft fuperfluous, that a frequent, an irregular, a fluttering and intermitting pulfe commonly indicates danger, fometimes approaching death : yet I muft add, that an intermitting pulfe fometimes attended the favourable crifis of a peculiar fpecies of fever. It was obferved, however, in fuch cafes, that the pulfe was not otherwife irregular, than by failing in its ftroke at the end of every third or fourth pulfation, neither was it generally found to be uncommonly frequent. Some inftances of this fingular appearance occurred to me during the time that I remained in Jamaica: fo that I was in fome degree difpofed to rank the intermitting pulfe among the figns of favourable crifis, in a fpecies of fever, the preceding courfe of which had been diftinguifhed by fymptoms of a peculiar nervous affection. When I became acquainted afterwards with the obfervations of Dr. Solano, I began to doubt whether the

L

intermiffion of pulfe, which I had met with in the
fevers of Jamaica, might not have been a fign of approaching diarrhea, which had not occurred to my
notice, rather than a fign of proper crifis, as I had
formerly imagined. I remained in this uncertainty
till lately, that fome inftances of this fymptom happening at the termination of fevers in this country,
have helped to confirm me in the opinion which I
entertained before. I found in thofe cafes to which
I allude, that the pulfe intermitted after every third
or fourth ftroke on the day, on which I expected
the crifis. The intermiffion of the pulfe was not of
fuch a nature as indicated approaching death; I therefore looked watchfully for a diarrhea, but no diarrhea
enfued. It muft be confeffed, indeed, that one of
the patients feemed to be much diftreffed with gripes
and flatus; but being deprived of the power of fpeech
we could not obtain any accurate idea of his feelings:
and no evacuation actually took place, till the day
following, before which time the intermiffion had difappeared altogether.—Befides the above, there are
fome other figns of pulfe which have their particular
indications; but they are fo generally known, that it
will not be neceffary to enlarge on the fubject. I
fhall therefore only obferve further, that changes from
better to worfe in the ftate of the pulfe, as the difeafe advances in its progrefs, are bad, while the oppofite changes are favourable: yet I muft likewife
add, that in thofe cafes of favourable change, it will
be neceffary to diftinguifh carefully the pulfe of coma,
from the pulfe of returning health.—The difference
is fometimes fcarcely to be known, except from the
attending fymptoms.

Next to the ftate of the pulfe, I fhall mention
thofe appearances of the tongue, which, together
with other concomitant circumftances, frequently
afford figns of the mildnefs or malignity of the difeafe.
Though we do not expect that the tongue fhould be

of a healthy afpect, during the continuance of a fever; yet where it is dry only in a moderate degree, or where it is covered with a fmooth and whitifh coat, the difeafe for the moft part, is void of malignity, though not always of danger. On the contrary, where it is immoderately dry, or dry and black, the indications of danger are great, and I may add, ftill greater where a white flimy and glutinous fubftance covers its furface. This flimy ftate of the tongue was often feen at an early period, and as far as my experience goes, conftantly indicated malignity. To the above we may add, a fodden or parboiled appearance of the tongue, which was not of lefs dangerous import than the preceding. But befides thofe obvioufly unhealthy afpects of the tongue, its appearance in fome inftances was not different from its natural ftate, except in a certain lividnefs of colour. This was conftantly fufpicious, and if not fatal, was always extremely dangerous. The danger indicated by the tongue, when it is intenfely dry, rough, cracked, or ulcerated is generally known; but I muft not omit to mention, that when from a dry and unhealthy ftate, it turns moift fuddenly, or affumes its natural appearance, whilft the other figns of favourable crifis did not fhew themfelves at the fame time, a change of the mode of action of the febrile caufe is indicated,—and generally a dangerous one.—I fay nothing of palenefs and tremor, as thefe fymptoms only indicate certain ftates of general or particular debility of the nervous fyftem.

Vomiting is another of the alarming, and fometimes of the dangerous fymptoms of the fevers of the Weft-Indies. If this fympiom continues during the remiffion of the fever, without material abatement, there is reafon to dread its conf·quences; but if it vanifhes or abates in a very material degree at the decline of the paroxyfm, it does not deferve to be fo particularly regarded. The practice however to

which it leads is often ferious. Vomiting is fuppofed by moft practitioners to indicate emetics; but
the indication is fallacious, and the practice is often
pernicious. During the time that I lived in Jamaica,
I had frequent opportunities of feeing vomitings rendered continual by the repeated ufe of emetics, which
before this treatment, appeared to be only accidental
fymptoms during the paroxyfm of the difeafe. I
therefore at laft became cautious of purfuing this
view, and have reafon to believe, that if I did
not oftener do good than others, I feldomer increafed
the danger. But befides the degree and frequency
of the vomiting, the nature of the matters thrown up
may likewife furnifh indications of the danger or
fafety of the fever. The various kinds of bilious vomitings have been fully explained, and the danger of
each has been fo particularly pointed out by many
writers, efpecially by Hippocrates, that I pafs over
the fubject without further notice, confidering it unneceffary to repeat the obfervations of others. I
muft however remark a more uncommon kind of
vomiting, which fometimes happened in the fevers
of Jamaica, and which I believe has hitherto efcaped
the notice of obfervers. The vomiting to which I
allude in this place, is a vomiting of a clear and
ropy liquor, in which are often found fwimming flakes
of a darker coloured mucus. This appearance was
chiefly obferved, where the remiffions were indiftinct,
and the fweats partial and incomplete. It conftantly
afforded an indication of danger, and I feldom found
that the ufual remedies were effectual in reftraining
it. Vomitings of black and vitiated matters are commonly known to be of the moft dangerous import,—
fucceeded by obfcure hickupings, they are often fore-
runners of death. Yet though this is generally true,
I muft not at the fame time omit to mention, that I
have feen feveral inftances of recovery where black
vomiting had prevailed for fome time; and other

cafes which give me room to conclude, that hickup-ings are not conftantly fatal. I take the prefent oppor-tunity to remark, that hickup was fometimes only a diftinguifhing fymptom of the difeafe, which increafed or declined with the paroxyfm; and that in other in-ftances it attended the favourable crifis of fevers, the preceding courfe of which had been characterifed by fymptoms of nervous affection. This fpecies of hickup was generally alarming in degree, and equally inexplicable with the intermitting pulfe, which I mentioned above as fometimes attending a favourable termination. It often continued the fpace of twenty-four hours, in fpite of all that could be done by me-dicine.

Next to the indications of vomiting, I fhall enu-merate fuch as may be drawn from the prefence or abfence of thirft. Immoderate and unquenchable thirft has always been reckoned an unfavourable fymptom in fevers. It is fo undoubtedly, yet I have frequently feen very extraordinary degrees of it con-tinuing for a length of time, without particular dan-ger. Befides the defire for liquid in general, there is often an unconquerable longing for drinks of a particular kind;—a feeling which ought always to be attended to, and frequently complied with.—The defire for cold water is fometimes ravenous.—I have known it not only fatiated with fafety, but even with good effects. But though this immoderate thirft is juftly reckoned a bad fymptom in fevers; yet an in-difference for liquid, with a dry tongue, and other marks of internal heat, is ftill worfe. It has indeed been generally confidered as fatal; but here we ought to diftinguifh, whether it proceeds from local affec-tion of the tongue and fauces, or from a general failure of the powers of life. In the one cafe it is a mortal fign, in the other it can only be faid to be dangerous.

The ancients, particularly Hippocrates and his

commentator Galen, have treated fo fully of the indications of evacuations downwards, that I fhould be able to do little more than to copy their obfervations. There is one fpecies of evacuation, however, which they do not appear to have defcribed very explicitly, and which I have often obferved to be attended with great danger. This is the frequent, fmall and ineffective excretion, and more particularly copious ftools, which refemble dirty water, efpecially if accompanied with tenfion of the hypochondria and abdomen.

Medical writers have been long accuftomed to form a prognoftic of the event of fevers, from puftular or fcabby eruptions about the mouth : but the fign is ambiguous, and cannot be depended upon, without many limitations. I fhall however relate that which has occurred to my own obfervation, without troubling myfelf about the opinions of others. And I remark in the firft place, that an eruption about the corners of the mouth, and near the lips, which comes forth freely, and foon turns into a fcab, particularly if it does not appear till after the third revolution of the difeafe, affords a general fign of fafety, at leaft it affords a fign that the complaint has attained the height of its violence. On the contrary, an eruption which fhews itfelf at an earlier period, which is crouded, and makes its way with difficulty, or which refembles iron-burnt blifters rather than puftules properly fo called, particularly if it is on the upper lip, and fpreads towards the nofe, affords a general indication of danger and malignancy.—Small and imperfect eruptions likewife are frequently a fign of a tedious difeafe.

The ftate of animal heat is another of thofe circumftances, which may be confidered as affording an indication of the nature and event of fevers. Where the heat of the body, in the remitting fever of Jamaica, was equally diffufed to the extremities, or not differing from an increafed degree of natural warmth, the difeafe was ufually mild, without particular danger or malignity ; but where acrid, fiery and pungent, though perhaps not much increafed in degree, danger was apprehended with reafon, particularly if the warmth was not extended equally to every part of the body. In the remiffions of thofe fevers, which were diftinguifhed by fymptoms of nervous affection, or, as is more commonly believed, putrefcent tendency, the heat of the body was often feveral degrees below the ftandard of health. The fymptom was alarming, but it was not in fact of much confequence. This diminution of the heat of the body, during the remiffion, was not by any means a rare occurrence; but befides this, there was fometimes obferved a degree of coldnefs, during the favourable crifis of nervous fevers, of a very fingular and extraordinary kind. In fome inftances this coldnefs was not inferior in degree to that of a perfon dying, or actually dead; yet a diftinction was perceived without difficulty. It was not accompanied with marks of ftricture on the furface of the body, at the fame time that the pulfe was generally foft, regular and full.

Next to the ftate of animal heat it will not be improper to confider the indications of the various kinds of fweats. The figns of a favourable fweat are commonly known. Where that excretion was fluid, warm and univerfal, particularly where accompanied with a foft, full and expanding pulfe, calm and eafy refpiration, general relief from fymptoms of diftrefs, with a cheerful eye and countenance, we might in general prefume on fafety of the difeafe, often on its favourable termination. On the contrary, where the fweat was cold, clammy and partial, particularly where the pulfe became or continued frequent, fmall and tenfe, with anxiety, reftleffnefs and difturbed refpiration, a circumfcribed flufhing, a greafy hue of the countenance, or a wild and dejected appearance

of the eye, the fituation was then alarming :—death,
in fhort, was feldom far off. There is, however, an
obfervation with regard to this fubject, which I muft
not omit to mention. Authors, with one confent,
have confidered cold fweats as certain mortal figns in
fevers ; but there appear to be exceptions to this general rule. I met with feveral inftances, while I remained in Jamaica, where univerfal fluid fweats, of
an extraordinary degree of coldnefs, accompanied the
crifis of the difeafe. I was much alarmed when this
appearance firft occurred to me, but my fears foon
vanifhed, as I found that the pulfe became flower and
fuller, that the refpiration became calm and eafy, and
particularly that the eye and countenance acquired
fuch a cheerfulnefs and ferenity, as are ufual at the favourable termination of fevers.

To the figns of prognoftic, which I have mentioned above, I fhall add thofe which are indicated by
the general ftate of the vital powers, or by the more
particular affection of parts, which are of immediate importance to life. Among the firft of the affections of the vital organs, we fhall confider fuch indications as arife from a difturbed ftate of refpiration.
A frequent, a hurried and unequal refpiration, (I do
not fpeak of that which depends on primary affection
of the lungs), is juftly confidered as a fign of a bad
difeafe. This is more certainly the cafe, where accompanied with deep and heavy fighing. Frequent
fighing was a common fymptom in the fevers of Jamaica, where the powers of life were depreffed; and
though not abfolutely a mortal fign, it conftantly indicated danger.

Next to the ftate of refpiration, I fhall mention the
ftate of the intellect, or reafoning faculty, which often
afforded fome prognoftic of the event of the fever.
Delirium, I obferved before, was a common fymptom
in the remitting fever of Jamaica. Where it vanifhed
or abated as the .paroxyfm declined, it was feldom

found to be of material confequence. On the contrary, where it continued during the remiffion undiminifhed in degree, it was a fymptom of the moft ferious nature. I mentioned in a former part of this treatife, that the caufe of fever appeared to act on the brain and nervous fyftem, in two general and oppofite ways; that is, by occafioning excitement or depreffion. Of thefe two modes of action, depreffion was the moft dangerous; unlefs where the excitement ran uncommonly high. But though I obferved, that there are only two general modes of operation, viz. excitement and depreffion; yet I muft alfo add, that the modifications are numerous, and very varioufly combined. Among the moft dangerous and alarming fpecies of the derangements of the intellect, we might reckon a ftern fullernefs, an unmanageable furioufnefs, picking the bed cloaths, tracing figures on the wall, and fuch other inftances of perverted judgment. Stupor and fufpenfion of the nervous influence, as we might term it, were likewife greatly to be dreaded : unlefs they fhewed themfelves only during the time of the paroxyfm, they were generally fatal, more certainly fo, if they followed convulfions.

As nearly connected with delirium, we fhall now confider other difturbed ftates of the functions of the brain, viz. the ftates of reft and watching. We do not expect that fleep fhould be found and undifturbed in fevers; yet we have been accuftomed to think favourably of the difeafe, where the patient is refrefhed by it. On the contrary, total want of reft, or unrefrefhing flumbers, conftantly indicate danger. There is, however, an aftonifhing diverfity of conftitution in this refpect, that muft always be taken into the account in forming an opinion. Want of fleep was obferved to give rife to delirium in fome perfons very fpeedily; others fupported it for a great length of time, without any appearance of delirium or fpafmodic affection. An appearance of

sleeping, without actually enjoying the comforts of
sleep is well known to be a dangerous symptom; yet
it is not by any means a mortal one. Anxiety and
restlessness, are often referred to the stomach; but
restlessness and jactitation, as depending on the state
of the nervous system, were likewise frequent, and
generally dangerous symptoms. Tremors of the
tongue and of the hands were common appearances
in fevers, with marks of nervous affection; but I
have likewise met with instances, where the whole
body shook, when any motion was attempted, not
otherwise than it does in paralysis or chorea sancti
viti. Startings and subsultus tendinum were not uncommon; and they were justly considered as indications of danger; sometimes as forerunners of con-
vulsion. A disposition to faint, even actual fainting, was frequent in the fevers of this country. It
was always dangerous, though perhaps less so, than
other symptoms which were less alarming, particularly if it suffered increase and diminution with the
paroxysms and remissions of the disease. But besides
these symptoms, which indicate diminished energies,
or irregular action of the nervous influence, we
may likewise observe, that the sphincter muscles frequently lose their power of contracting, particularly
in the advanced periods of fevers. Thus persons
sometimes can only lie upon their back, the eyes and
mouth are half open, the powers of speech and swal-
lowing are impaired or lost, and urine and stools pass
off without consciousness or against their will. It is
unnecessary to remark, that these are all symptoms
of the most extreme danger. If they proceed from a
general and uniform diminution of the powers of
life, we may justly consider them as fatal; if they
are only produced by a certain mode of action of the
febrile cause, and are remarkably increased during the
paroxysm, or aggravated by peculiar circumstances
of constitution, we shall find many instances of re-

covery. Thus, I have often feen people recover, who could neither fpeak nor fwallow; who did not appear even to diftinguifh objects, and who were unable to retain their urine and ftools; or who were not confcious when they paffed; yet I do not pretend to have met with any of thefe fortunate events, where thefe alarming fymptoms were the confequence of uniform diminution, or general extinction of the vital principle.

From the figns which I have enumerated, feparately and collectively confidered, we may in general be able to form fome prognoftic of the nature and event of the fevers of Jamaica. If to thefe we add thofe indications, which may be taken from the ftate of the eye and countenance, we may attain a ftill more fatisfactory knowledge. It is an obfervation which I have conftantly found to be true, that where the eye and countenance were ferene and cheerful, the difeafe was void of any latent malignity, though it might be otherwife of a dangerous nature. On the contrary, where the appearance of the eye was fad, watery, inflamed, or uncommonly gliftening; where the countenance was of a dreary hue, downcaft, dark and clouded; and fometimes where it was of a beautiful blooming colour, which was not natural to the patient, there was always reafon to fufpect danger. But though a ferene and cheerful eye and countenance are generally indications of fafety, I muft not at the fame time omit to mention, that it fometimes happens in beginning mortifications, or in imperfect or unfavourable crifis, that the eye and countenance affume, for a fhort time, this flattering appearance of ferenity and compofure, though the hour of death is actually approaching faft. The indications from the eye and countenance are of the greateft importance, in enabling us to form a judgment of the event of fevers; but little of this knowledge is, in fact, communicable in words. It muft be drawn, in a great meafure, from our own obfervations.

I do not pretend that the figns of prognoftic, which
I have enumerated in the preceding pages, are by any
means complete, if referred to fevers in general; but
I at the fame time believe, that they are lefs defective, if applied more directly to the remitting fever
of Jamaica. They were collected at a time when I
was not much acquainted with books : and, on that
account, I am induced to offer them to the public
with more confidence; particularly, as I find that
the indications, to which I have principally trufted,
appeared in the fame light to fome authors, who are
univerfally confidered as careful obfervers, and who
practifed in climates, in many refpects, fimilar to that
of the Weft Indies. I have difcriminated, as far as
was in my power, between the doubtful and more
certain appearances of danger or fafety ; and I hope
I have no where advanced any thing, which has a
tendency to miflead the uninformed. I may add, that
general knowledge in prognoftic goes no farther than
a very rude outline, which individuals muft fill up
from their own experience. There are, in fact, few
figns in fevers, which are abfolutely decifive in themfelves ; and as thefe figns are often varioufly combined, fo they muft be feparately and collectively ef-
timated. It is only from confidering accurately the
refult of the whole, that we can be enabled to fpeak
with confidence.

# C H A P. VIII.

IT is certainly a matter of some importance, to be
able to distinguish between actual crisis and the
simple remission of the fever of Jamaica; but it is a
matter about which the practitioners of that country
did not seem much to concern themselves. A difference undoubtedly exists, and the marks of it appeared
to me clear and unequivocal in most instances. It
was otherwise in the intermitting fever of America.
In the pure intermitting fever of that country, I sometimes guessed luckily; but I cannot say positively,
that I ever discovered signs on which I could depend
with absolute certainty, that the fever was gone, not
to return again, till the hour of return was past. In
this disease, on the contrary, I should not expect to
be deceived once in a hundred times. Much of this
information, however, is too minute to be intelligi-
bly explained in words, and therefore can only be acquired by actual observation.

I shall attempt to enumerate the principal of those
signs, from which we may be enabled to attain some
knowledge of the difference between actual crisis and
temporary remission : and in the first place I remark,
that the tongue was usually rough and dry, even during the most perfect remissions of the endemic fever
of Jamaica. If it therefore happened, that it assumed a smooth and moist appearance at the end of a paroxysm, there generally was reason to believe that the
fever was gone, not to return again. This presumption was still stronger, where its edges acquired the
cherry-like colour of health ; and particularly where
the coat, with which it was usually covered, shewed

M

a difpofition to loofen and feparate. It muft however be remarked, that though thefe appearances of the tongue afforded a common mark of the termination of fevers, they did not by any means afford a decifive one. Inftead of crifis, they fometimes only indicated a change in the mode of action of the febrile caufe; they were, in fhort, in fome cafes only forerunners of fymptoms of nervous affection, or marks of a change from a continued to a remitting or intermitting form. Such are the prefumptions of actual crifis, which might be drawn from the appearances of the tongue, where the tongue happened to be remarkably changed from its natural appearance in the preceding courfe of the difeafe. But it alfo fometimes happened in cafes of the fingle tertian, where the paroxyfms were flight, and the remiffions long and perfect, that the tongue was fo little altered by the prefence of the fever, as not to afford any certain criterion between the remiffion and actual crifis. In fome fevers likewife of a malignant kind, the tongue was fometimes fmooth and moift, even red and clear on the edges, whilft the difeafe was advancing rapidly. This, however, fo far as I have feen was conftantly connected with a particular ftate of the ftomach, viz. with naufea, or with vomitting of a vifcous liquor.

Signs of crifis taken from the pulfe alone, were not in general much to be depended upon in the fevers of Jamaica; yet, together with other circumftances, the ftate of the pulfe might often help us to decide in doubtful fituations. Changes from bad to better, if no fymptoms of comatofe affection appeared at the fame time, were generally confidered as indicating crifis, or tendency to crifis; yet it will be lefs expected, perhaps, that I fhould rank the intermitting pulfe among the figns, which indicate a favourable termination of the difeafe. Some inftances of this have occured to me, both in the Weft-Indies and in England; but though I mention the fact, I confefs myfelf unable to offer an explanation of it.

The state of the skin affords marks more decisive of the total or temporary absence of fever, than the ordinary state of the pulse. When the sweat, which in the preceding remissions had been partial and imperfect, became copious, fluid, universal and of long continuance, there was generally a presumption of crisis. But independent of the nature of the sweat, there is something in the state of the skin, something in the impression which it makes on the hand which feels it, very different when the fever has only remitted, and when it has terminated finally. Though it may be difficult, perhaps impossible, to mark this distinction precisely in words ; yet it is easily known to those, who accustom themselves to observe minute circumstances with attention. There was, in fact, no indication of that spasmodic stricture on the surface, which had been observable in the former remissions, and we might say, that the body was perspirable, even in the extreme parts. It is a circumstance likewise not a little curious, that the heat of the body, during a crisis, particularly in those fevers, which had principally affected the nervous system, was sometimes so much below what it usually is in health, as to be really alarming. In some instances, I have found the extremities to be not less cold, than if the patient had been actually dead; yet this coldness was of such a nature, as to be distinguished without much difficulty from that which precedes death.

The above is only a very imperfect history of those signs, which actually distinguish the crisis of fever from a simple remission. There are still some others, not less to be depended upon, perhaps, but which cannot be so easily reduced to distinct description.—— Among these, we may reckon unusual evacuations upwards or downwards, sound and refreshing sleep, where watchfulness had prevailed through the preceding course of the disease, return of natural appetites, decrease of thirst, loosening of scabby eruptions, and

M 2

above all, a certain expreffion of chearfulnefs in the
eye and countenance, which though not to be defined
in words, conveys to the mind of the obferver, a
ftrong conviction of what is going forward. This
brightnefs of the eye was well known to Hippocrates, as a falutary fign in fevers; but though it generally affords a very decided indication of a favourable
event; yet we muft be careful to diftinguifh from it
that clear and glaffy appearance, which the eye fometimes acquires previous to death.

# CHAP. IX.

BEFORE I attempt to offer rules for the treatment of the remitting fever of Jamaica, it will be proper to confider in the firft place, how far the cure of the difeafe is the work of nature, and how far it already has been, or hereafter may be accomplifhed by the exertions of art. The queftion is important, and till its limits are defined, we cannot hope to eftablifh rules of practice on a firm bafis, or to conduct a mode of treatment on a confiftent plan. It will be neceffary however, before entering farther into this fubject, to confider the power of the vis medicatrix naturæ :—a principle, which under one name or other, has influenced the views of medical men from the earlieft records of phyfic, till the prefent times.

---

# SECTION I.

## OF THE VIS MEDICATRIX NATURÆ.

IT is an opinion, which feems either to have been exprefsly avowed, or tacitly acknowledged in every age of the world, that a fever is a combat or effort of nature, to remove from the fyftem the derangements of a morbid caufe; or in other words, to reftore a difeafed body to its ordinary health. It was long believed, that the powers of the conftitution made an attempt to concoct the crude and undigefted humours;—and finally to expel them from the body. But this mode of reafoning is found to be unfatisfac-

tory: and fome late authors have adopted the idea,
that nature directs her efforts towards effecting a folution of fpafm, on the prefence and obftinacy of
which the difeafe has been thought, in fome meafure,
to depend.  It will be a hard tafk to reconcile this
difference of opinion, or to fay which is the right
one.  There are plaufible arguments on both fides
of the queftion; and neither fuppofition, perhaps,
is accurately true.  The coction of humours (as it
may be called) is often apparently connected with
evident approaches towards a relaxation of fpafmodic
ftricture; and increafed difcharges by the different excretories, are obvioufly attendants of its actual folution.  Yet though this is certainly true, it ftill is not
clear, that either the coction of the humours, or the
folution of the fpafm, is the real effect of a regular
mechanic operation of the powers of nature, attempting by this means to overcome the deftructive tendency of the difeafe.  It is even more probable, that
the coction of the humours, or folution of the fpafm
are only circumftances of accident, occafionally connected with certain ftates of action of the morbid
caufe; but which do not arife from the regular defign
of nature to accomplifh this purpofe.  I juft now obferved, that there is a difference of opinion about the
mode of operation, which nature employs to combat
the effects of the difeafe; but I may add, that no
body, as far as I know, attempts to deny the exiftence
of fuch a principle in the conftitution of the frame,
as difpofes it to reftore its own health by a certain
train of regular efforts.  On the firft view of the
fubject, indeed, there are many circumftances which
give countenance to the opinion.  The relief which
often follows hemorrhage, fweat and other evacuations
in unufual quantity; and ftill more, the relief, which
attends the eruption of the fmall pox, or the appearance of gout on the extremities, undoubtedly affords
a ftrong prefumptive argument, that nature raifes

fome active and generally well directed efforts, which remove from the body a caufe that difturbs the ordinary functions of health. But though the above circumftances are commonly known, and may be fuppofed to afford an argument in fupport of this opinion; yet the fact may perhaps admit of another explanation, while there are other appearances, connected with the fubject, which render the exiftence of this regular defign of nature very queftionable. When I fay, that I cannot readily allow the vis medicatrix naturæ, (in the fenfe in which it is generally underftood), to be an eftablifhed principle in the conftitution of the frame, I am aware, that I incur an imputation of ☞ prefumption. An opinion, venerable from antiquity, and fupported by many plaufible arguments, might be thought to be fecure from the attacks of a man, who has no profeffional reputation to boaft of: yet as no defire of novelty has induced me to fabricate a conjecture on a dark fubject, fo I humbly hope I may be indulged in my attempt to explain an important truth. The fanction of two thoufand years, and the authority of the names who fupport this doctrine, are formidable opponents; yet I truft I may be able to prove, that the vis medicatrix naturæ does not, as is commonly believed, reftore the health of the body by one general and uniform mode of operation; or that our difeafes are not removed in confequence of a regular defign in the mechanifm of the frame. I fhall relate the fact from which I have been led to form this conclufion; for the refutation or confirmation of which, I require no other indulgence than a candid enquiry.

I have juft now declared, that fever, or the caufe of fever, is not combated and finally overcome by a regular train of active efforts, or a vis medicatrix naturæ: and I muft obferve in proof of it, that there were many of the fevers of the Weft-Indies, where the difeafe, or the paroxyfm of the difeafe, vanifhed or declined, without any exertion on the part of nature being perceiveable. The powers of life, during this period, were in fome degree fufpended. The patient, who could only be faid not to be actually d ad, was totally infenfible to every object that was near to him; and often did not feel the irritation of acrid fubftances that were applied to him: yet after a certain conti- nuance in this ftate, he began to refume his powers of fenfation and motion; and the difeafe at laft fub- fided or vanifhed, though the efforts of nature were not difcovered; nay, though the vital powers were fometimes in a ftate fo weakened, as to be apparently very little capable of effort. This fact, which the moft fuperficial obferver could not eafily overlook, furnifhes fufficient reafon for doubting of the very exiftence of a vis medicatrix naturæ; a doubt which is further corroborated, by obferving the manner in which death frequently approaches in the fevers of Jamaica. It is known, that the fufferings of the pa- tient are fometimes alleviated for a fhort time before death. This alleviation, wherever it does take place, happens conftantly at the expected period of crifis. The caufe of it has generally been attributed to the vis medicatrix naturæ; that is, to a laft effort of the powers of life; but I have weighed carefully all the circumftances connected with the phenomenon, and cannot readily affent to the opinion. To indulge in conjectures, is contrary to the principles I profefs; yet I muft fuggeft, that a fever, or the paroxyfm of a fever, terminates, ftrictly fpeaking, from a hidden fomething in the nature of the febrile caufe, from fomething which ceafes to act, or which changes its mode of action after a certain duration. I do not pretend to explain the manner in which this happens. I only add, that the fact is fupported by probabilities. It was frequently obferved, in fuch cafes of fever as terminated fatally, that there was actually a period of time, generally the period immediately preceding a decided fatal termination, where the real prefence of difeafe was perceived with difficulty. The patient, however, was unable to recover. Death happened in a very few hours, and it feemed to enfue in fuch cafes, from one of the following caufes, viz. either from the mechanifm of a part of vital importance being deftroyed; from the powers of nature being too much exhaufted to continue life; or perhaps ftill oftener, from the recurrence of the difeafe, in another form, fpeedily putting a period to exiftence, while the vital principle was in this weakened ftate. But though the circumftances, which I have mentioned, afford grounds for believing that a fever, or the paroxyfn of a fever, is not actually removed from the body, folely by the efforts of a vis medicatrix naturæ; yet if we choofe to proceed further in the inveftigation, it will be no difficult tafk to involve this opinion. which has hitherto been confidered almoft as facred, in ftill greater perplexity. If we admit of the exiftence of a vis medicatrix naturæ, it will not be eafy to conceive, how a fever, which has once been expelled from the body, fhould return again in a given time, or how the alternate paroxyfms of the double tertian, for inftance, fhould be of fuch different duration or of fuch different degrees of violence in the fame perfon; neither can we underftand, how a fever of one kind fhould laft only feven days, another fourteen, and another twenty, or longer:—circumftances which happen daily, without the leaft apparent connexion with the innate vigour of conftitution. We fhall be equally puzzled likewife, if we attempt to explain on the fuppofition of this principle, how a fever fhould continue, while the powers of the conftitution are vigorous and ftrong, and ceafe when they are apparently exhaufted. The above are well known facts, and do not leave any room to doubt, that the termination of fever, or of the paroxyfin of fever, depends on fome other principle befides the mere efforts of the vis medicatrix naturæ. Whether this refides, as was hinted before, in a hidden modification of the nature of the caufe, which ceafes to act, or changes its mode of action after a certain duration; or whether, combined with this, the conftant but imperceptible changes, which are continually going on in the fyftem, deftroy in the frame of the individual, that particular ftate of aptitude to the febrile caufe, in which the difeafe confifts, we cannot determine with certainty; yet it would be obftinate to maintain any longer, that the cure of fever is owing to general and well directed efforts of nature, expelling a morbific matter or overcoming a prevailing fpafm. It is true, that an obvious folution of fpafmodic ftricture, or the appearance of a morbid matter on the lefs important parts of the body, ufually attend the favourable termination of fevers; yet thefe circumftances are, in fact, attendants rather than caufes of crifis. I do not deny, that increafed difcharges by different outlets, fometimes moderate the violence of fever during its continuance, as well as attend its final folution; yet it has not, nor perhaps can it ever be demonftrated, that this proceeds from a regular defign of nature.

The hints which have been thrown out in the preceding pages, give room for fuppofing that the vis medicatrix naturæ, in the fenfe in which it is ufually underftood by medical writers, is only a principle of doubtful exiftence in the conftitution of the frame; yet though this is certainly true, I do not pretend to deny, that the animal machine is endued with a power, which refifts, in fome meafure, the derangements of a deftroying caufe, and which perfifts to a certain degree in continuing the action of living. The general nature of the caufe of fever, or the nature of its various modifications is a myftery, which we d not as yet know. We only know, that when  fent in a certain ftate of vigour and activity, it d    s or difturbs the actions and functions of the fyftem;—

while we likewife know, that it does not always difturb every action or every function in the fame degree. It has occurred too often to have efcaped the
moft fuperficial obfervation, that where one part of
the body fuffers particularly, the others are often relieved in proportion. We frequently in this manner
obferve, that general fever is diminifhed by the appearance of local pain; or, on the contrary, increafed by its removal. It likewife often happens
from the fame principle, that where the ftomach and
biliary fyftem fuffer much, there is lefs diforder in the
other parts: and on the other hand, that where thefe
fufferings are removed or mitigated, the general fever
runs higher, and often continues high, till the fame,
or other local affections, are again produced. Thus,
though we are totally ignorant of the intimate nature
of the caufe of fever, we ftill perceive very plainly,
that it either poffeffes fomething in its own nature,
or accidentally meets with fomething in the conftitution of the individual, which determines it to affect
the different parts of the body in an unequal degree.
It ufually exerts its greateft force upon parts, which
are preternaturally weakened by the general influence of climate, feafon, fituation, or other accidental
caufes. Hence bilious appearances are common in
the hot months of hot climates, pneumonic affection
in cold and dry weather, greater degrees of vafcular
excitement among the temperate and more active
races of men; while fymptoms of nervous affection
prevail among the luxurious and enfeebled. The
above, with other fpecies of the increafed action of
the caufe of fever on a particular part of the body,
depend wholly, perhaps, on circumftances of accident; yet it has fo happened, that thofe irregular determinations have unfortunately been confidered as
the efforts, which nature employs to expel from the
body a caufe, which difturbs the economy of health.
I fhall not, at prefent, go fo far as to contend, that

thefe determinations are not, in fact, intentions of
nature; but fhall only beg leave to fuggeft, if they
actually are intentions, that it is mere chance which
determines whether they are falutary or fatal.  It is a
truth which nobody will deny, where the force of the
difeafe is accidentally directed to an organ of excretion, or to a part of little importance to life, that the
reft of the body is often proportionally relieved, and
even that a recovery of general health is fometimes
the confequence; yet the contrary is the effect, where
the functions of the part, upon which the force of the
fever has been thus accidentally diverted, are of immediate importance to the action of living.  The
Gout, a difeafe, the caufe of which bears no very remote analogy to the ca fe of fever, may be adduced
as affording an illuftration of this truth.  The proximate caufe of gout, is equally hid from us as the
proximate caufe of fever.  We know, however, that
the one equally with the other, has a tendency to deftroy life.  We likewife know, that there is a power
or principle in the conftitution, which to a certain
degree refifts deftruction.  The nature of this power,
however, is unknown.  We are not only in the dark
with regard to its nature; but we can only form conjectures about the part where it principally refides.—
We, however, clearly perceive its force and activity
to be different in different parts of the body.  We
may next be allowed to remark, that where the caufe
of gout is in a certain ftate of modification, tumults,
(which properly enough may be termed re-action),
arife in the fyftem, and go on to continue till this
caufe or hurtful matter finds an outlet from the body,
or a lodgment on one particular part.  The outlets
from the body are numerous: the parts on which the
gout feems principally to fix its feat, are the extremities, where the power of refiftance is fmalleft.—
The vital principle, however, becomes weaker as
man advances in years; and the caufe of the diforder

feems then frequently to find accommodations in parts, which are lefs remote from the fources of life. This more efpecially is the cafe, where tone and vigour have been preternaturally weakened. Hence the ftomach, the bowels, fometimes the brain, and even the heart itfelf fuffer from the immediate action of this difeafe, in the latter periods of life. But though no perfon perhaps will deny, that the caufe of gout finds readieft accommodation, (if I may fo apply the term) in thofe parts of the body, where the vital powers are naturally weak, or have been accidentally weakened from various caufes; yet we may add, that it is likewife removed from the parts, on which it has been thus fixed, by fuch applications as excite their active powers; or, in other words, which call forth the local re-action of the fyftem. We may alfo obferve, that tumults arife in the general fyftem, in confequence of this repreffion or repulfion of the morbid caufe from a particular part; and that they do not in general ceafe, till an outlet is opened, or accommodation found in fome other parts of the body. The above appearances, occur daily in the hiftory of gout. They feem to bear a ftrong analogy to thofe irregular determinations, which frequently take place in fevers, and their caufe perhaps is the fame. We do not perceive any other law by which they can be explained, than the natural or adventitious ftate of activity of the powers of life, which refift deftruction with unequal force in the different parts of the fyftem: fo that we fhall be obliged to conclude, that thofe fufferings, which have hitherto been ftyled the efforts of nature, are in reality more of the paffive, than of the active kind.

The circumftances which I have now mentioned, combat the very exiftence of the opinion, which has been commonly received with regard to the vis medicatrix naturae. I have hinted, that the extent and limits of that principle are narrow, and that the fa-

N

lutary effects are accidental. I shall next endeavour
to shew, that they cannot, without danger, be made
the basis of the general plan of cure in febrile diseases. The task is important, but the attempt may
be thought presumptuous, as an opinion, contrary to
that which I advance, has obtained almost the universal consent of mankind. I have no desire of
changing names, or of making distinctions, where
there is in fact no difference. I perfectly acquiesce
in retaining the word vis medicatrix naturæ, provided it is limitted to a certain mode of re-action, or
to a power in the constitution of resisting destruction
unequally in its different parts, in consequence of
which, irregular determinations sometimes prove sa-
lutary by accident; yet I must add, that if we mean
to denote by this term a system of laws, which have
the best directed tendency to remove from the body a
cause which destroys health, and endangers life, the
opinion has a very uncertain foundation. There are
few persons so ignorant, or so blindly devoted to the
doctrines of αυτοκρατεια as not to own, that the usually reputed efforts of nature, are often ill directed,
sometimes pernicious : in short, that they are obvi-
ously the causes of death. The truth of this observation cannot be denied, and unfortunately it obliges
the advocates of the vis medicatrix naturæ, to grant
the conclusion, that the laws of the principle are imperfect. The works of the author of nature, as far
as our limited knowledge can trace them, are univer-
sally without defect, if examined according to the plan
on which they have been originally formed. If they
appear otherwise, it becomes us to hesitate, before
we decide. We may not have comprehended the
fundamental principle of the design; but we revolt
from the idea, that the execution would be left imperfect, had it been intended by the Author of our
being, that the mechanism of the frame would be
such, as should oppose and remove, in the most

effectual manner, the derangements of the morbid cause. Defect and imperfection can have no place in the works of the Almighty. Had it actually been in the original design of our Creator, that the human body should be provided with a system of the best concerted laws for restoring its health, when deranged by the numerous causes of difeafes, as it is impious to suppofe, that thofe laws could be defective; fo we may reafonably conclude, that the effects of fevers would not then have been fatal. We find however, that fevers, as well as other difeafes, are fatal to people of all ages and defcriptions: and that nature's intentions of cure, if they really are intentions, are often deftructive to herfelf. I need fcarcely remind the reader of examples of their pernicious tendency. Vomitting, fweating, increafed difcharges by ftool, &c. are generally confidered as the falutary efforts of nature: but inftances are numerous, where the excefs of thofe evacuations have obvioufly proved the caufes of death. In the fame manner, abfceffes, which in the remote parts of the body, fometimes attend, and even fometimes perhaps influence the favourable termination of fevers; in the brain, or in other organs of importance, are no lefs certainly the caufe which deftroys life. In both inftances the defign of nature, if it can be called a defign, is the fame. The force of the difeafe being turned principally upon one part, the reft of the body is in a great meafure relieved from its fufferings ; - but the health and ftructure of the part are hurt or deftroyed by the change; and it depends wholly upon the accidental importance of the organ, upon which this diverfion has been made, whether death or recovery is the confequence. Thus it often happens, that the reputed indications of nature prove the immediate caufes which deftroy the exiftence of the individual ; a fact not reconcileable, with the infinite power and wifdom of the Author of our being.

I have infinuated, that the efforts of nature are uncertain and precarious. They depend on accidental determinations to different parts of the body; and I may add, that if we endeavour to inveftigate the caufe, which directs the mechanifm of the frame, to adopt one fpecies of effort, or one mode of determination in preference to another, we fhall not per⸗ haps be able to find any other, than a difference in the ftates of the powers of life, which refift deftruction with unequal degrees of force in the different parts of the body. Where there is the leaft refiftance, either from the natural or accidental circumftances of the conftitution, there the difeafe moft obvioufly exerts its greateft force. Hence we are fufficiently warranted to conclude, that though the ftructure of the human body is perfect with refpect to every purpofe for which it is intended, being only endued with a principle, which refifts deftruction, or perfifts in continuing life to a certain degree; yet that it is extremely defective, if we confider it as a machine furnifhed with a fyftem of laws, which have an invariable and well directed tendency to reftore health by the moft judicious and rational efforts. The reftoration of health, in confequence of this re-action, or irregular determination which takes place in the fyftem, is only a circumftance of accident. The fkill of man fometimes fucceeds, where the efforts of nature have obvioufly failed.

## SECTION II.

### OF THE GENERAL INDICATIONS OF CURE IN FEVERS.

THE vis medicatrix naturæ, has been hitherto efteemed a principle of much importance in the cure of febrile difeafes. I have attempted to ex-

plain its real limits and extent ; but am afraid I may
not have done it fatisfactorily. A tumult which,
properly enough perhaps, may be termed a reaction
of the fyftem, evidently takes place in confequence
of the application of a morbid caufe ; but there feems
to be little reafon for believing, that this reaction
points out the beft method of cure, or wholly by
itfelf accomplifhes this important bufinefs. But
though the reputed efforts of nature are thus defective in accomplifhing the cure of fevers ; yet I do not
deny, that there is a general tendency in fevers, or in
the paroxyfm of fevers, to terminate in a given
time, often by a fixed and regular mode of termination. We do not, however, by any means comprehend the caufe upon which this depends. From the
fimilarity in the progrefs and termination of epidemics,
as well as from the fteadinefs with which various
forts of fevers purfue their courfe, in fpite of the
moft oppofite modes of treatment, we are led to conclude, that there is fomething peculiar in the modification of the caufe, which influences the duration of
the difeafe. This at prefent, is unknown ; perhaps
is a knowledge which we cannot attain ; yet if we
take pains to obferve the courfe of fevers with attention, we may difcover fome rules of practical ufe.
We know that one fpecies of febrile difeafes, obfti-
nately purfues its courfe, notwithftanding every endeavour to oppofe it ; while another is fo totally under
our management, as to be ftopt fhort at pleafure
with almoft infallible certainty. It thus happens,
that the intermittent is perfectly under our controul.
Over the continued, and even over the remitting
fever of Jamaica, I am afraid, we fhall be obliged to
confefs, that we poffefs no very certain power.—But
I fhall examine this fubject more particularly.

When I firft arrived in Jamaica, in the year
1774, I found that the practitioners of that country
very generally believed, that the courfe of the ordi-

nary endemic fever was checked with great certainty
by the powers of Peruvian bark. This opinion, indeed, is found in every medical book, and it appeared
frequently, on the firft view of the fubject, to be
well founded. No great fpace of time, however,
elapfed before fome circumftances were obferved,
which prefented the matter in a different light. I
found in many inftances, that bark was given in the
firft remiffion, or on the fecond day of the difeafe;
in others, it was not given till the third remiffion, or
till the fixth day from the beginning of the complaint;
and in fome again, the fever difappeared altogether
before a fingle grain of this remedy had been adminiftered. I was particularly exact in marking the
time or the period of the difeafe, at which the bark
was begun to be given, as alfo the quantity which
was taken upon the whole. The refult was not fuch
as might have been expected. Notwithftanding the
moft oppofite modes of treatment, the difeafe appeared to terminate or change about the fame periods
in almoft every patient. This fact was confirmed in
numerous inftances; and it feems to afford a very unequivocal proof, that bark, in the quantity in which
it is commonly prefcribed in the Weft Indies, has
not the effects which are *ufually* afcribed to it. I do
not, however, infer, that this remedy may not be capable, with more decifive modes of management, of
effecting all that has been expected from it. I had
not, during the time that I remained in Jamaica, any
conception that the ftomach could have retained, or
that it would have been fafe to have ventured upon
the quantities of bark, which I afterwards gave to
others, or took myfelf in the intermitting fever of
America. Two fcruples or a drachm, every two hours,
is in fact only a fmall dofe. To this under dofe,
during fhort remiffions, we might perhaps impute the
failure of that remedy in the fevers of the Weft
Indies. That this actually is the cafe, is confirmed

in fome degree, by an inftance which I find recorded among my notes. A young man was feized with a fever, about the middle of Auguft, which fhewed marks of great violence from the beginning. Bark was given early, and in larger quantity than cuftomary. The laft paroxyfm of the difeafe, was in fome meafure fufpended, in confequence of this proceeding ; yet, except that the marks of external fever were obfcure, the patient remained, as ufual, uneafy and diftreffed, till the period at which the crifis was expected ; when the marks of final termination fhewed themfelves diftinctly. This is the only cafe I met with, where the paroxyfms of the fever of Jamaica were ftopt, or fufpended by the bark; or where external marks of fever vanifhed without evident figns of crifis. It affords only a doubtful proof of the power, which this remedy has been fuppofed to poffefs, of abfolutely cutting fhort the courfe of the endemic of that country. But though the bark was feldom efficacious in abruptly cutting fhort the courfe of this difeafe, it is no more than juftice to remark, that it is a remedy which was almoft every where fafe, and that it was ultimately ufeful in promoting the cure. It imparted in moft inftances, where it was employed, a degree of tone and vigour to the fyftem—a certain fomething to the conftitution ; in confequence of which, the crifis, which we fhould have expected to be only partial or imperfect, became decided and final. I have fuggefted thofe few remarks, with regard to the virtues of Peruvian bark in the common endemic of Jamaica ; if its effects are fo very doubtful in this difeafe, we have no reafon to expect, that they will be more certain in fevers of a more continued kind.

Befides bark, the power of which appears to be very precarious, other remedies have been employed by phyficians, with the view of cutting fhort the courfe of fevers.—Antimony, under one form or

other, has been celebrated for this intention, since its
first introduction into medicine. James's Powder
is the most famous, and perhaps the most effectual
antimonial preparation, which, as yet, has been offered to the public. I am sorry to say, that I had
not an opportunity of making proper trial of it, in
the fevers of the West-Indies ; but I can add, that
the emetic tartar was often found to be dangerous,
scarcely ever effectual in cutting short the course of
the disease, unless given at an early period, or before
the fever had assumed a proper form. Its *virtues*, as
a febrifuge, were heightened by the addition of opium
and camphire, I am sorry also to remark, that I cannot speak with confidence of the virtues of James's
Powder, in the intermitting fever of America. Emetic tartar was frequently employed, but it did not by
any means answer the expectations which were entertained of it. I acknowledge, that it might be so
managed, as apparently to prevent the return of a
particular paroxysm ; but the instances, where it completely removed the disease, were so rare, that I do
not consider it as possessed of very eminent virtues.
I have had frequent opportunities, since my return
to Britain, of trying James's Powder in the continued fevers of this country ; and the result of my
experience inclines me to believe, that this remedy,
when given at an early period, has sometimes actually carried off the disease. It appeared likewise,
when exhibited near the critical periods, to render
the crisis more complete ; but I have little cause for
thinking, that it ever cut short a fever in the midst
of its course. Thus it appears, that these two celebrated remedies,—bark and the various preparations
of antimony, are, in fact, less effectual in cutting
short the course of febrile diseases, than has been commonly supposed ; I cannot, however, abandon the
idea that the purpose, which has been expected from
these remedies, may still be obtained by other means.

Galen mentions fome inftances, where he extinguifhed the fever by copious evacuations : authors mention many, and I have myfelf feen fome, where drinking plentifully of the coldeft water, produced the fame effects. The alternate ufe of warm and cold bathing occafions great changes in the ftate of the conftitution : and from the trials, which I have made of thefe applications, I do not entertain a doubt, that they may be fo managed, as to fhorten very materially the duration of feyers. I do not however promife, that they are capable of being fo conducted, as infallibly at once to ftop the difeafe in its progrefs. This can only be accomplifhed by thofe great and remarkable changes, which deftroy a certain aptitude, in the ftate of the fyftem, to the morbid caufe, in which the difeafe is fuppofed to confift. But I muft at the fame time confefs, that as we neither know the nature of this aptitude, nor the particular nature of remote caufes, fo every attempt of cure on this plan, as it muft be at random, cannot be adopted without danger. It is a view, therefore, which will not be profecuted with fafety, while our knowledge of the nature of morbid caufes, and of the laws and ftructure of the human frame, is fo imperfect.

It is evident from the facts which I have related in the preceding pages, that we cannot fafely trufl the cure of fevers to thofe tumults, or irregular determinations in the fyftem, which are ufually ftyled the efforts of nature : neither does it appear, that we can depend on the efficacy of any one remedy, we are yet acquainted with, as poffeffed of the power of abruptly cutting fhort their courfe. We ftill however perceive, that thefe difeafes have a general tendency to terminate in a given time and fteadily to go through a regular progrefs, in fpite of the greateft exertions of art. If we review the practice which medical people have followed in fevers, from the days of Hippocrates to the prefent times, we meet with fuch contradictory methods of treatment, as render

it impoffible to avoid pronouncing, that if one man
had actually faved life, another's endeavours feemed
as if intended to deftroy it: yet few authors have
ventured to offer the fruits of their labours to the
public, without previoufly boafting more fuccefsful
methods of cure, than were known to their predeceffors. Hence, if we are not fometimes difpofed to
doubt their veracity, we can hardly avoid concluding,
that their practice had been feeble, and of fmall effect.
We lament, with reafon, that medical facts are frequently of little value: nay, that they oftener miflead,
than guide us in the way to truth. An overfondnefs
for ourfelves is, perhaps, more the caufe of this, than
real want of candour; the natural propenfity of the
human mind to flatter itfelf, difpofing us to attribute
cures to remedies, which were adminiftered near the
critical periods of the difeafe; while twenty inftances,
where fimilar treatment produced no apparent effect,
are infenfibly blotted from the memory. This at
leaft was the cafe with myfelf. I flattered myfelf in
many inftances, that I had actually faved life :—
I now find, on maturer reflection, that I had in reality
done no material good. Thus it frequently happens,
I believe, that practitioners boaft of cures, to which
they have no right; at the fame time I am convinced,
that they are frequently charged with deaths, of
which they are innocent. The life of man does not
appear to depend upon fo fmall a matter, in febrile
difeafes, as is generally imagined; and is not often
preferved or endangered by the routine of common
practice. It is not always eafy to know exactly the
real effects of treatment; neither has the road, which
leads to this knowledge, been purfued with fufficient
induftry. Fevers occur frequently, and on that very
account, have been traced lefs minutely in their courfe,
than fome other difeafes. There are few practitioners,
who write down in the prefence of the fick, a minute
and accurate hiftory of the various cafes of fevers, which come under their care; who obferve

carefully the changes which happen from day to day;
who note the particular methods of cure, and the effects which arife in confequence of every alteration
of treatment.   Yet, unlefs all thefe circumftances are
attentively confidered, we cannot hope to form conclufions, which are in any degree to be depended
upon.   If we defer making remarks ʼtill the patient
recovers or dies, difficulties will be eafily got over,
and fuch facts as contradict opinions, in which we
have long believed, will be more eafily reconciled,
as being lefs perfectly remembered.   Hence it is
that a man may continue a very extenfive practice,
for a very long life, without ever once getting a
view of the real truth.

As from what has been faid before, little doubt
can remain of the precarioufnefs of trufting the
cure of fevers to the fimple efforts of nature ; or, if
we except intermittents, to the efficacy of particular
remedies which cut off their courfe abruptly, it remains
to look out for fome other principle, which may ferve
us as a guide in the conduct of our practice. It is a truth
to which we may accede without hefitation, that the
caufe of fever, as I have mentioned before, whatever
it is, or however modified, has a general tendency to
deftroy the powers of life ; while we likewife know,
that there is a principle in the mechanifm of the frame,
which refifts d ftruction to a certain degree.   We
alfo know, that the caufe of the difeafe differs in degree of force ; and that it is differently modified according to various circumftances ; as we likewife obferve, that the principle of life, or power of refiftance,
is different in activity, in the different parts of the
body.   To thefe two powers, viz. the force of the
morbid caufe and the powers of the conftitution our
views in practice muft be principally directed.
Hence we may eftablifh a general rule, that whereʼer
the force of the morbid caufe is weak ; at the fame
time that the powers of life in the general fyftem, and
particularly in the organs of vital importance, are

ftrong and active, we have nothing more to do than
to look on. On the contrary, where the caufe of
the difeafe is of unufual force, or where the powers
of life are preternaturally weak, our interference ought
to be fpeedy, bold and decided. In other words, we
leave the bufinefs chiefly to nature, or take it entirely
out of her hands. It is this which a practitioner muft
firft determine, when called upon to give his affiftance
to a patient labouring under a fever. If the powers
of the conftitution appear to be equal to the tafk,
our interference would be officious, and perhaps might
prove hurtful. If affiftance is neceffary, it ought to
be our principal aim to render it complete ; but in
profecuting this view, we meet with much difficulty
and impediment. We are not yet acquainted with
any one remedy, which has a certain and infallible
power of cutting fhort fevers in the midft of their
courfe. It is not therefore abfolutely in our power
to take the bufinefs entirely out of the hands of nature.
We can, in fact, go no farther, than to oppofe her
pernicious efforts ; or to obviate the fatal tendencies
of the difeafe. The fatal tendencies of the difeafe,
are varioufly modified, and the means by which they
muft be obviated, are fometimes directly oppofite.
Two general modes, however, of the fatal action of
fevers may in moft inftances be difcovered. The
caufe of the difeafe, in one cafe, exerts its influence
on the fources of life and motion ; in the other, the
ftructure of an organ of lefs importance is deftroyed,
and death happens only from a fecondary effect.
There is perhaps no fundamental difference in thefe
different modes of action; yet the indications of cure,
which arife from this view, are totally oppofite. In
the firft inftance, it is neceffary to excite, and to fupport the general powers of life : an indication of very
great extent. In the other, it is fometimes neceffary
to diminifh the general reaction of the fyftem ; to
obviate irregular determination, and to oppofe with
vigour the tumultuary efforts of nature.

# CHAP. X.

I SHALL begin this fubject with obferving, that the
fever, which prevailed in the diftrict of Savanna
la Mar, was naturally a difeafe of the remitting kind;
yet circumftances were fometimes connected with it,
in fuch a manner as prevented it from affuming its
proper form. To remove thofe circumftances, which
thus mafked or concealed the real genius of the
difeafe, was confidered as the firft ftep towards a cure.
The accomplifhment of this purpofe, however, was
fometimes difficult; neither could it always be effected by the fame means. Thus it happened frequently
in cafes, where there was excefs of excitement, or a
high degree of inflammatory diathefis, that the remiffions where fcarcely perceptible; as it was likewife
obferved, that where there was a want of reaction, the
paroxyfms were often languid and obfcure. In the
one cafe, the remiffi ns difc vered themfelves in con-
feq ence of bleeding, dilution and copious evacuation;
in the other, wine and cordials determined the difeafe
to affume its proper genuine form.

In the firft place, evacuations were ufually employ d as the means of procuring remiffion, where
the inflammatory diathefis prevailed in excefs; I may
add, that they were prope for the moft part, and that
they feldom failed of producing the effect. Bleeding
was frequently neceffary, and generally of fervice. Its
efficacy, however, was often heightened by particular
modes of management. Thus relaxation of fpafms, and

O

removal of inflammatory diathefis, more certainly follow ed bleeding, if the blood was drawn from a large orifice; if the patient was placed in an erect pofture, during the operation; and mo.e certainly ftill, if the lower extremities were at the fame time immerfed in warm water. When bleeding had been premifed, and repeated acco ding to the circumftances and urgency of the cafe, it was then cuftoma.y to open the body freely: for which purpofe, I have not found any thing anfwer better, than a thin folution of Glauber or Epfom Salts, with a fmall portion of emetic tartar. The operation of this medicine was extenfive. It migh. be fo managed as to eccafion naufea, or moderate vomiting; to operate brifkly downwards, or to prom te a gentle diaphorefis. Remiffions were gene.ally the confequence of this method o proceeding, whe e there was no defect in the manner of conducting it. But where it it fo happened, that the circumftances of the patient forbad the ufe of this l xative; or where it might not be proper to carry it to a fufficien. length, benefit was derived from a powder, compofed of ni re, camphire, emetic tartar and opium, given in pre ty large do es, and repeated frequently. Remiffion, at leaft a great abatement in the violence of fymptoms, was generally the confequence of this plan o. treatment; particulai , if affifted by the plentiful dilution of watery liquors, by warm bathing and by large glyfters of fimple water. It is fuperfluous to mention the ufe of blifters in cafes o local affection; but it will be lefs expected, that this remedy fhould be recommended in fevers, where there is an excefs of the general inflammatory diathefis.—I can, however, bear teftimony to its efficacy. The manner by which blifters produce their effects, is not yet agreed upon among authors; neither do I pretend to throw any new light upon the fubject; but I would beg leave to fuggeft, that the mode of affording relief in the prefent, at leaft, did not feem to be much

unlike the effect of local affections, in confequence
of which the violence of fevers is fometimes obferved
to fubfide.

I purfued the above method of procuring remiffion
in thofe fevers, where there was real inflammatory
diathefis prevailing in excefs ; but it fo happened, that
the figns of this diathefis were fallacious, appearing in fome inftances to be prefent, though the real
genius of the difeafe was actually of a different nature; a circumftance, which occafioned a difference
of management in conducting the method of cure.
Exceffive evacuations were not only unfafe in f ch
cafes, but in general had not any powerful effects in
difpofing the difeafe to affume a remiting form.
Bleeding, however, was often found to be neceffary,
though it was feldom requifite to repeat the operation. The good effects which were ob erved to follow the ufe of cathartics, were not in general very
remarkable ; yet it was proper, in moft inftances, to
open the body freel, ; for which purpofe, no form of
medicine, with which I am acquainted, anfwered
better than a folution of falts with a fmall portion of
emetic tartar, and fometimes with the addition of laudanum. In cafes of local pain, blifters applied near
the feat of affection were always of eminent fervice ;
and in cafes of general irritability, they were often
equally ufeful, when applied to the back part of the
head and neck. A powder compofed of nitre, camphire, emetic tartar and opium, was likewife employed with fuccefs ; but the liberal ufe of warm bathing, was ftill more to be depended upon. No perfon, perhaps, will refufe confent to the method of proceeding, which I have hitherto recommended ; but
when I mention a free and bold ufe of cold bathing,
even in an early ftage of this fever, I do not expect
the fame conceffion. To dafh cold water on the
head and fhoulders of a perfon in a fever, has an appearance of rafhnefs and hazard. I can, however,

O 2

produce the teftimony of repeated experience for the fafety of the practice, no lefs than for its fucce's in procuring remiffion; and fhall therefore confider it a duty to recommend it warmly to the public. Wherever it was employed,—and the cafes in which it was tried were numerous, a cal n and equable perfpiration, additional tone and vigour, with great abatement of irritabilit, were conftantly obferved to enfue.

The paroxyfm' and remiffions were generally diftinct in the beginning of fevers, where the nervous fyftem was principally affected; but often became lefs fo, as the difeaf: advanced in its progrefs; a circumftance which did not arife oftener from the natu e of the complaint, than from the common method of treatment. Bleeding was often difpen'ed with in the fevers of the Weft-Indies; but vomiting and purging were indulged in with freedom. The diftinction of paroxyfm and remiffion was fometimes evidently rendered obfcure by this practice; while it was likewife obvioufly refto ed again, by the ufe of wine and cordials, which excited the powers of life. In this fpecies of difeafe, evacuations were feldom neceffary; feldom indeed adviffible in a great extent. Bleeding unlefs under particular circumftances, was totally impr per. Cathartics were fometimes dangerous, and antimonial vomits often funk the patient irrecoverably. Bliftering, on the contrary, even at an early period, was generally of fervice; as alfo were opiates, and a judicious ufe of the warm bath; but cold bathing with falt water, was, of all others, the remedy of the moft powerful effect. I do not pretend to fay, that it abfolutely ftopped the courfe of the fever; but I can fay with truth, that it generally reftored the diftinction of paroxyfm and remiffion, diminifhed irritability, and imparted a degree of tone and vigour to the fyftem, which was juftly confidered as a fign of fafety.

To procure remiffion in fever, diftinguifhed by a prevalence of the putrefcent tendency, is not in every inftance an eafy tafk. A remitting fever, with marks of fpecific putrefaction, is not a difeafe of common occurrence in Jamaica; but a fever with figns of putrefactive tendency, mixed with fymptoms of great irritability, or a high degree of malignity, is not altogether rare. From he complicated nature of the diforder, the indications of cure are often difficult and perplexed. Bleeding is univerfally condemned; more, I believe, from theory than from actual obfervation. It was, and perhaps ftill is, a fafhionable mode of reafoning, to impute the languors and other marks of debility, which are common in the fevers of the Weft-Indies, to a putrecent tendency in the fyftem. Such fymptoms however are in fact more generally the attendants, or diftinguifhing figns of fevers, where the nervous fyftem is affected. In fuch cafes, bleeding is obvioufly hurtful; in the one of which we now treat, (where fuch a difeafe actually exifts,) it is not only a remedy of fafety, but of very eminent fervice, previous to the application of cold, particularly previous to cold bathing, which may be ufed with freedom and boldnefs. Cold bathing, indeed, is the remedy on which we muft principally depend. There are others which do good occafionally; but this is the only one I know, which has any very confiderable effect in changing the nature of the difeafe. There is a general rule in the practice of medicine, which requires to be particulaly remembered in thofe complicated fpecies of fever, viz. that as the indications of cure are often embarraffed; fo the appearances, which pincipally point to danger, are firft to be attended to; while the plan of cure, which we determine to be the moft proper, muft be followed up with vigour and refolution. We ought always to bear in mind, that in dangerous and difficult cafes feeble remedies, or even powerful ones timidly ufed, are of little avail. Cold bathing, employed with timidity, failed of doing good in fome inftances. I

O 3

met with no example, where the boldeſt uſe of it did harm. It was ſeldom, I muſt again repeat, that it did not ſucceed in obviating irritability, in checking the putreſcent tendency, and in imparting to the ſyſtem that degree of tone and vigour, in which ſafety is obſerved to conſiſt.

The method of procuring remiſſion, in thoſe fevers which were diſtinguiſhed by local affections, or irregular determinations to particular parts, was nice, and ſometimes difficult. Bleeding was frequently proper, eſpecially, if there ſubſiſted at the ſame time marks of a general inflammatory diatheſis : but it was ſeldom ſufficient wholly to accompliſh the buſineſs. However, together with a judicious management of warm bathing, it greatly heightened the good effects of bliſters, the remedy on which the principal dependence was placed. In fevers which were accompanied with uncommon pain of the head, I have ſometimes found it ſerviceable to apply cold to the part affected ; the feet being at the ſame time immerſed in warm water, and blood flowing by a large orifice from the arm. I alſo frequently obſerved, that the general fever ran higher, though it likewiſe more certainly aſſumed its proper form, in conſequence of bleeding, bliſtering, and the removal of local pain. In thoſe fevers, where bilious appearances were the effect of accidental, irregular determination to the ſtomach or liver, the remiſſions were often obſcure : neither did the method of treatment, which was generally adopted, ſeem to be well calculated to bring forth the natural, genuine appearance of the diſeaſe. Bilious appearances, it muſt be owned, ſometimes vaniſhed, while the type of the fever became more diſtinct after the exhibition of an emetic or briſk cathartic ; yet there is cauſe to doubt if this depended on the evacuation of bile. It might be ſaid, with more truth, perhaps, that the action of the emetic, by exciting the powers of the ſtomach and biliary ſyſtem, effected a change in the irregular determination, which had formerly taken place to thofe parts. It was generally obferved, where good effects did not follow the firft exhibition of remedies of this kind, that harm was ufually the confequence of a fecond. Vomiting, in fhort, was often rendered continual, and the diftinction of paroxyfm and remiffion was apparently deftroyed, in confequence of the operation of violent emetics. Inftructed by repeated examples of their hurtful effects, I at laft fcarcely ever employed antimonial vomits; even the fafeft kinds were ufed with caution. If it appeared, at any time, that the action of vomiting would be ferviceable, camomile tea, or at fartheft a few grains of ipecacuana were generally thought fufficient for the purpofe. When this bufinefs was finifhed, a draught of cordial ftimulating liquor, which had a tendency to promote a diaphorefis, was next adminiftered. By this mode of treatment, efpecially if a blifter was applied at the fame time to the region of the liver, I have the fatisfaction to add, that the bilious appearances for the moft part vanifhed, and, if care was taken to fupport a determination to the furface, feldom ever returned during the continuance of the fever. Different feafons, and different fituations of country were particularly diftinguifhed by correfponding determinations. Thus a tendency to the bowels and biliary fyftem was chiefly remarkable in the autumnal months, and in low and champaign countries; the head and breaft were oftener affected in the winter months, and in hilly fituations.

I obferved before, that it is the firft object in the cure of fevers, to remove thofe circumftances, or accidental ftates of the body, which hinder the difeafe from affuming its proper form. Thus, to procure remiffion, appeared univerfally to be the firft bufinefs in the cure of the fever of Jamaica; the next, and a very important one, is to prevent the return of the paroxyfm. If we knew a remedy, which could be depended upon to accomplish this purpose with certainty, the cure of the disease would be easy; but the Peruvian bark, which almost infallibly stops the course of intermitting fevers in all countries, does not seem so indisputably to possess the same power over the usual endemic of the West Indies. I am sorry to own, that my knowledge on this subject, is not altogether satisfactory. At the time I left Jamaica I did not entertain a doubt, that the endemic of the district, where I resided, possessed something in its own nature which decidedly determined its duration. It was usually observed to terminate on a critical day, generally by very evident signs of crisis, and without seeming to be materially effected in its course, by the various and opposite modes of treatment, which were sometimes pursued. But though this was true for the most part, yet the superior efficacy of the very large doses of bark, which I afterwards ventured to give in the intermitting fevers of North America, obliges me to speak with hesitation, when I mention the powers of that remedy. It is probable that bark, with the same management, might have had the same effects, in the fever which prevailed at Savanna la Mar, as in that which is commonly epidemic in Georgia and the Carolinas; yet no doubt remains, that as commonly employed in Jamaica, it has no right to be considered as a remedy, which abruptly cuts short the course of the disease. In every case where it was tried, (except one) it did not seem to do more than give a degree of tone and vigour to the system, to excite a certain state of inflammatory dia-thesis, in consequence of which, the crisis was observed to be more perfect and complete, though it did not perhaps actually happen at an earlier period. Thus I am disposed to conclude, from all the experience which I have had, that bark is not generally carried farther, in the cure of the remitting fever of Jamaica, than merely to support the tone and vigour of the powers of life. If we truft to it for more, in dofes of two-fcruples or a drachm given every two hours, we fhall certainly be deceived.

Having mentioned the different methods of treatment, by which it was attempted to procure remiffion in the endemic fever of Jamaica, and having likewife endeavoured to afcertain how far we can go in preventing the return of paroxyfms, it only remains to detail fome particulars in the management of the plan of cure, where the different fpecies of fevers were diftinguifhed by a peculiar train of fymptoms. It was obferved in general, that fevers, with a moderate degree of inflammatory diathefis, feldom required our interference. The difeafe, after a certain duration, terminated ufually of its own accord. After I had gained fome experience of the general courfe of fevers, I ufually allowed thofe, in which I did not perceive marks of danger, to go on their own way, that I might better difcover thofe periods, at which the difeafe was naturally difpofed to terminate. Thus where the paroxyfms continued regular and diftinct, the remiffions perfect, and the vigour unimpaired, nothing material was attempted to be done. On the contrary, where the paroxyfms were long, or lefs diftinctly formed, with figns which indicated an approaching affection of the nervous fyftem, bark, and other remedies, which excited and fupported the powers of life, were given with the earlieft opportunity. Changes from inflammatory diathefis to nervous affection, were obferved to happen frequently on the fifth day. Bark, and fuch remedies as imparted tone and vigour to the fyftem, were given without delay; and the difeafe terminated for the moft part on the ninth. In thofe fevers, which were of a complicated nature, in which figns of inflammatory, nervous, or putrid diathefis were varioufly mixed, blifters applied in different manners, opiates, bathing and antifpafmodics were often materially ufeful; but it would be arrogance to attempt to defcribe rules for the particular mode of application, which muft vary more or lefs in almoft every cafe, and which only can be learned from actual obfervation. There is one rule, however, in the treatment of fevers, of which the practitioner ought never to lofe fight, viz. that wherever it is neceffary to interfere, it is only the moft vigorous decifion which can do good. We cannot, as was faid before, depend with certainty upon bark, as a remedy poffeffed of the power of abfolutely cutting fhort the caufe of the fever of Jamaica; yet wherever the fevers of that country difcovered figns of nervous affection, I do not know any thing in the materia medica, from which fuch beneficial effects may be expected. If it did not actually ftop the difeafe, it was eminently ferviceable in conducting it to a favourable iffue. Opium, wine, fnake-root, &c. were often obferved to heighten its good qualities : but the particular ufe of fuch additions can only be regulated by circumftances. Wine has been freely recommended in fevers with fymptoms of nervous affection ; and it muft be owned, that its good effects were confiderable, not only in real debility, but wherever the caufe of the difeafe acted by weakening or depreffing the powers of life. Wine was likewife obferved to be more ufeful in cafes of mobility and weaknefs, than in cafes of ftupor and fufpenfion of the nervous influence. But though it is actually a remedy of great value, its virtues appear to have been greatly enhanced. In many inftances it was not proper in any quantity ; in fome, it was only proper in a fmall quantity, and in very few, perhaps, could we allow of the quantities which are given in common practice. At one time I carried the ufe of wine in the nervous fever of Jamaica, to a very great length ; but I afterwards learnt, that a third part of the quantity would have probably anfwered the purpofe better. Though it undoubtedly is an ufeful cordial and tonic, it is ftill inferior to cool air, and particularly to cold bathing.

In thofe treacherous and malignant fevers, which I have defcribed in the third fection of the fixth chapter, the ordinary medical aids were often feeble and infufficient. The ccurfe of the difeafe was generally rapid. There was little time left for deliberation; and where the firft, at leaft where the fecond remiffion paffed over, without fome bold and decided fteps to prevent the return of the paroxyfm, or change the nature of the fymptoms, the opportunity was probably never again in our power. The type of this malignant fever was ufually of the fingle tertian kind; yet it was often found to anticipate, by fuch long anticipations, that the fecond paroxyfm fometimes made its appearance, before any decided fteps were taken by the practitioner to ftop its progrefs, or often, indeed, before there were any furmifes of danger. As this difeafe feemed to have a nearer refemblance to the intermittent, than fome other fpecies of the endemic of Jamaica; fo bark appeared to be capable of producing more effect, in the prefent inftance, than in thofe cafes where the remiffions were more obfcure. It ftill, however, deferves to be remarked, that fuch fcanty dofes, as were ufually given in Jamaica, could not be effectual; indeed, we could fcarcely hope, that any powerful alteration could arife from lefs than half an ounce given at the fhort interval of every other hour. I confefs candidly, that I never ventured fo far; yet I alfo confefs, that I am by no means fatisfied with the fuccefs of the method of cure, which I adopted at firft, in this difeafe. I loft fome patients before I was aware of danger, and perhaps fuffered others to die, from a dread of ftepping over the bounds of common practice. At laft, I acted with more decifion; and have caufe to be fatisfied with the fuccefs of the attempt. As foon as I difcovered the malignity of the difeafe, the marks of which, for the moft part, were plain in the courfe of the firft paroxyfm, the head was immediately fhaved, and covered with a blifter, which reached half way down the neck; the feet were likewife put into warm water, previous to the expected return of the paroxyfm; the body was rubbed with brandy or rum; wine, and fometimes opium were given in quantity fufficient to exhilirate the fpirits, or to produce a low degree of intoxication; bark was alfo prefcribed in larger dofes than ufual, during the remiffions; and cold bathing was occafionally employed with freedom. I do not fay, that the return of the paroxyfm was abfolutely prevented by this method of treatment, but I have the fatisfaction to add, that the fatal tendency of the difeafe was evidently obviated by it.

I muft farther obferve, that bark has been confidered as the principal remedy in thofe fevers, the nature of which has been believed to be putrid; but the great fame of this remedy has proceeded from theory, rather than from actual obfervation. A real putrid difeafe, (as I have faid before), if we except the yellow fever, occurs very feldom in Jamaica. Symptoms of languor and debility, however, are frequent in the fevers of that country; appearances which, though in fact, only figns of nervous affection, have often been imputed to a putrid tendency in the habit. Bark has been found to be efficacious in thofe cafes of languor and debility, which have been falfely thought to arife from a putrefactive tendency in the fyftem. Hence this remedy has gained credit, on a foundation which does not exift. I may add with truth, that the power of the bark was very equivocal, in thofe cafes where the marks of putrid tendency were obvious. The principal dependence was then conftantly placed in the cold falt-water bath; which, if well managed, produced moft aftonifhing effects. An additional quantity of falt was often added to the water of the fea; and care was taken

that the coldnefs fhould be as great as the circumftances of the climate would permit. The chief dependence I have faid, was conftantly placed in this remedy, where the putrid tendency prevailed in the
general fyftem; but where it was more particularly
confined to the bounds of the alimentary canal, faline
draughts, in the ftate of effervefcence, were ufeful,
as were likewife glyfters of cold water impregnated
with fixed air. It is fuperfluous to mention cool air,
clean linen, cold drink and the liberal ufe of wine.

The remedies, which I have hitherto mentioned,
are fuch, as feem chiefly intended to fupport and in-
vigorate the general powers of life; but which are
lefs immediately directed to obviate the fatal tendency
of particular fymptoms; which I propofed to confider, as the fecond indication of cure. I fhall probably depart materially from the ufual ideas of practitioners in the profecution of this fubject. I do not
deny, that bark may be given with fafety in fevers,
which are accompanied with local affections, or irregular determinations to particular parts; yet I muft
likewife obferve, that bark is not the remedy on
which the weight of the cure depends. Local affections, or irregular determinations are often diftinguifhed by the name of the efforts of nature. I do
not indeed difpute, where the part of the body, to
which the determination takes place, is of little
importance to life, or more certainly, where it is an
organ of excretion, that fuch parts as are of greater
confequence, are, in fome degree relieved in confequence of this effect; and though there is ftill an impropriety, there is lefs danger in confidering this irregular action of the morbid caufe, as an effort of nature, or a quality of the vis naturæ medicatrix, by
means of which, the health of the fyftem is eventually rendered more fecure. This, however, is only
a circumftance of accident. It does not appear to
depend upon a regular defign of nature, and cannot

be admitted with fafety as the bafis of a plan of cure.
We find, in fhort, that though the life of the whole
is fometimes preferved by it, the deftruction or derangement of the part is generally the confequence.
But that the fallacious tendency of thofe tumultuary
efforts of nature may be more clearly illuftrated, I
fhall adduce the example of fevers, diftinguifhed by
an increafed fecretion of bile. It is an opinion,
which feems to date its origin from Hippocrates, that
bile vitiated in quality, or redundant in quantity,
deferves to be confidered as the caufe of the fpecies
of difeafe diftinguifhed by the name of ardent fever :
and it muft be confeffed, that the frequent appearance
of bilious difcharges, in the fevers of hot climates,
gives countenance to the fuppofition. From the frequency of this fymptom, perhaps, the practitioners of
the Weft-Indies adopted the idea, that bile is the caufe
of the fever of that country ; while the method of
cure, which they ufually purfue, has ferved to confirm them in their error. Prepoffeffed with an opinion
of the prevalence of bile, they adminifter cathartics
and emetics with a liberal hand. If bile appears in the
firft evacuations, they confider it fufficient authority to
proceed; if it does not appear, they conclude that the
remedy has not been of fufficient force to reach the
feat of the difeafe ; and therefore perfift in their intentions, till the effect is at laft produced. It is well
known, that a repetition of cathartics and emetics
feldom fails to produce the appearances of a bilious
difeafe. Hence this fymptom of fever, and all the
dangers which follow it, are frequently the work of
our own hands. That this is the cafe, appears from
a relation of the method of cure, which I ufually
adopted in fuch fevers as were diftinguifhed by fymptoms of this nature at an early period. Inftead of
encouraging the vomiting, or promoting the evacuation of bile downwards, I generally did every thing
in my power to moderate, or even to check it.—

Sometimes I prefcribed an emetic; but it was more with a view to excite the action of the ftomach and biliary fyftem, than to promote an evacuation of redundant or vitiated humours. After the operation of vomiting was finifhed, a blifter was ufually applied to the region of the liver, and fuch a plan of cure was purfued, as fupported a determination to the furface of the body, and gave tone and vigour to the ftomach and general powers of life. By this mode of treatment, bilious appearances vanifhed fpeedily, or ceafed to be troublefome ; while by the repeated ufe of emetics and cathartics, they generally continued long, and often prevailed throughout the courfe of the difeafe. We may thus, I hope, conclude, without any unnatural inference, that there appears to be danger in encouraging thofe tumults, which have been ufually confidered as the efforts of nature. If they are in fact efforts, no perfon can pretend to deny, that they are generally precarious; nay, that they are often the immediate caufes of death.

I have now detailed the particular fteps of the method of cure, which I adopted in the remitting fever of Jamaica ; a difeafe which I treated, in fome refpects, on a different plan, and if felf-love hath not blinded me with more fuccefs than the generality of thofe practitioners whom I had the opportunity of knowing, I treated the difeafe with fuccefs; but I dare not affirm, as fome have done, that under this method of treatment, I never loft a patient. I proceeded, indeed, with diffidence and diftruft of the powers of the medical art ; venturing no farther than to fupport the general powers of life, and to obviate fymptoms of a fatal tendency. Many pretend to cut fhort the courfe of fevers, by the force of a fingle remedy; but the means do not appear very obvious, and the effect was often precarious. I grant, that it is fometimes in the power of the practitioner to exterminate the caufe of difeafe by forcible means, or to deftroy a certain aptitude of conflitution, in which this difeafe may be faid to confift; but I muft at the fame time obferve, that there is danger likewife, leaft he extinguifh life. The bark, which has been fo much celebrated for checking the courfe of fevers, though generally fafe, is feldom effectual : others are frequently dangerous. During the time that I remained in the Weft Ihdies, I obferved attentively the ftate of body, which ufually attended recovery; as likewife thofe appearances which preceded, and apparently were the caufes of death. Tone and vigour, or a moderate degree of the ftate of body diftinguifhed by the name of inflammatory diathefis, without local affection, afforded the fureft figns of fafety ; general failure of the powers of life, or irregular determinations to organs of importance, were the moft certain appearances of danger. Thus, after obviating particular fymptoms of a fatal tendency, it was the principal indication to fupport the general powers of life, or to excite the tone and vigour of the fyftem. This was beft accomplifhed by bark, wine, cool air, and above all, by cold bathing, which I am induced to confider as the moft important remedy in the cure of the fevers of the Weft-Indies; and, perhaps, in the cure of the fevers of all hot climates. Though it might not abfolutely cut fhort the courfe of the difeafe; yet it feldom failed to change the fatal tendency of its nature.

# CHAP. XI.

THE difeafe, known in the Weft Indies by the name of yellow fever, particularly that fpecies of the difeafe diftinguifhed by black vomiting, has not, fo far as I know, been defcribed by the practitioners of any other country. In the autumnal fevers of moft climates, indeed, as well as in the remitting fever of Jamaica, yellownefs is not by any means uncommon: neither is vomiting of matter of a dark colour altogether rare, in the moments preceding death : yet in as much as I can judge, from what I have feen myfelf, or heard from others, the complaint which is the fubject of the prefent treatife, poffefes fome characteriftics of its own, different from thofe of every other difeafe. I do not pretend to determine, in what this characteriftic difference precifely confifts : yet I may fay with confidence, that the fpecies of this difeafe, which terminates in black vomiting, may be diftinguifhed with certainty from the autumnal fever of aguifh countries, or from the endemic remitting fever of Jamaica, even in the firft hours of its attack. I muft, however, at the fame time own, that there is a fpecies of difeafe, where the remiffions are obfcure, and where figns of nervous affection, or even fometimes of apparent inflammation, are more confpicuous than fymptoms of putrefcency, that I have fome difficulty in clafing properly. There is not any thing more foreign to my intention than multiplying names, or eftablifhing diftinctions which do not exift in reality; yet, as I have often feen inftances of fever, to which yellownefs and black vomiting did not feem to be effential; but in which paroxyfms and remiffions were extremely obfcure,

P 3

or perhaps did not exist, I at last ventured to conclude, that this species of illness had more affinity with the disorder usually known by the name of yellow fever, than with the common remitting endemic of the country. I shall therefore describe it as a species of that disease; though I am less solicitous about fixing its place in nosological arrangement, than of giving a history of it, by which it may be easily recognized.

I am not ignorant that there are several authors, who have written on the subject of the yellow fever; yet I had not the opportunity of consulting any of them, except Dr. Hillary, during the time that I lived in Jamaica. This writer's method of cure was adopted very generally by the medical people of the island, and many of them were disposed to think favourably of its success. I cannot however conceal, that I was disappointed in every instance where I saw it employed. If, in any case, the patient recovered, this fortunate event appeared to be more owing to great natural strength of constitution, or to a lower degree of disease, than to the efficacy of the method of treatment. But besides, that this author's method of cure is feeble and ineffectual, the historical part is particularly defective; the disease, in short, being only very imperfectly discriminated from the common endemic of that country. The consideration of these circumstances, induced me to throw together the observations which I had made on this subject, during the time that I resided in the West Indies; and though I am conscious that they are imperfect, I still hope that they contain some hints which may be useful to those, who have had little experience of the fevers of hot climates.

In our enquiries into the history of the yellow fever, some circumstances present themselves to our observation which are not a little curious. It has never been observed that a negro, immediately from

the coaft of Africa, has been attacked with this difeafe; neither have Creoles, who have lived conftantly
in their native country, ever been known to fuffer
from it: yet Creoles or Africans, who have travelled
to Europe, or the higher latitudes of America, are
not by any means exempted from it, when they return
to the iflands of the Weft Indies. Europeans, males
particularly, fuffer from it foon after their arrival in
the tropical countries; yet, after the natives of Europe
have remained for a year or two in thofe hot climates,
efpecially after they have experienced the ordinary
endemic of the country, the appearance of the yellow
fever is obferved to be only a rare occurrence. But
befides, that this difeafe feldom difcovers itfelf among
thofe people, who have lived any length of time in a
tropical country, it has likewife fcarcely ever been
known to attack the fame perfon twice, unlefs accidentally after his return from a colder region. The
remitting fever, on the contrary, does not ceafe to
attack fuch as have refided the greateft part of their
life in thofe climates; or who have lived after the
moft regular and abftemious manner; a fact which
feems to prove, that there actually exifts fome eflen-
tial difference between the two difeafes; or which
fhews, at leaft, that the revolution of a feafon or two
deftroys in the European conftitution, a certain aptitude or difpofition for the one difeafe, which it ftill
retains for the other.

Having thus premifed fome circumftances, which
regard the general nature of the yellow fever, I fhall
proceed to give a more particular defcription of the
difeafe, previoufly dividing it into three forms, in each
of which, I believe, I have frequently feen it appear.
1. Into a fpecies of difeafe, in which figns of putrefaction are evident at a very early ftage, which is
generally rapid in its courfe, and which ufually terminates in black vomiting. Yellownefs feldom or
never fails to make its appearance in the prefent in-

ſtance ; and perhaps it is only this form, which, ſtrictly
ſpeaking, can be called the yellow fever. 2. Into a
form of fever, which either has no remiſſions, or remiſſions which are ſcarcely perceptible ; in which
ſigns of nervous affection are more obvious than
ſymptoms of putreſcency ; and in which yellowneſs
and black vomiting are rare occurrences. 3. Into
another form, in which regular paroxyſms and remiſſions cannot be traced; but in which there are
marks of violent irritation, and appearances of inflammatory diatheſis in the earlier ſtage, which give
way after a ſhort continuance to ſigns of debility and
putreſcency, to which yellowneſs frequently ſucceeds,
or even ſometimes the ſo much dreaded vomiting of
matter of a dark colour. The diſcaſe, which I have
divided in the above manner into three diſtinct forms,
appears to be in reality only one and the ſame. The
difference of the ſymptoms probably ariſes from very
trivial or very accidental cauſes ; and it is a matter
of great difficulty to diſcriminate thoſe ſigns, which
are eſſential and neceſſary to its exiſtence. It is in
ſome degree peculiar to ſtrangers from colder regions
ſoon ofter their arrival in the Weſt Indies, and may
generally be diſtinguiſhed from the remitting endemic
of the country, not only by the obſcureneſs, or total
want of paroxyms and remiſſions, but likewiſe by a
certain expreſſion of the eye and countenance, with
ſomething unuſually diſagreeable in the feelings, of
which words convey only an imperfect idea.

## SECTION I.

I SHALL deſcribe, in the firſt place, the moſt
common and moſt formidable ſpecies of this diſcaſe, which, as I obſerved before, is diſtinguiſhed
by early ſigns of putreſcency, by an intenſe degree

of yellowneſs ; and, towards its termination, uſually by vomiting of matter of a dark colour. It was mentioned in the preceding treatiſe, concerning the remitting fever of Jamaica, that fevers of different types had their different hours of invaſion ; but no ſuch property was obſerved in the preſent diſeaſe. In ſome inſtances the yellow fever began in the morning, though the evening, upon the whole, was the more uſual time of its attack. The firſt ſymptoms were languor, debility and head-ach, together with an affection of the ſtomach peculiarly diſagreeable. This laſt often preceded the others, and was in ſome meaſure characteriſtic; but it is impoſſible to give a clear idea of it in words :—anxiety, nauſea, and certain unuſual feelings were ſo ſtrangely combined, that any deſcription, which I might attempt to give of this complicated ſenſation, would hardly be intelligible. The horror and ſhivering, which ſo uſually precede fevers, was ſeldom great in degree in the preſent inſtance ; but it ſometime continued long, and was often accompanied with ſenſations of a very unpleaſant kind. The heat of the body, though rarely intenſe, was frequently of an acrid and pungent nature. The pulſe was weak and confined in its ſtroke. It was likewiſe frequent, and the nature of the arterial pulſations were creeping or vermicular; in ſhort there was a perpetual motion under the finger, combined with ſomething, which gave the idea, that the diſeaſe was not of the kind which has paroxyſms and remiſſions. Together with this, the eye was ſad and watery; or in ſome degree inflamed, having much that appearance, which is the conſequence of expoſure to the ſmoke of green wood. The face was often fluſhed ; yet the fluſhing, in the preſent caſe, was different from that which ariſes from ordinary cauſes. There was a degree of confuſion, and often a degree of grimneſs joined with it, difficult to be deſcribed in words ; but which a perſon, ac-

quainted with the appearances of the difeafe, immediately recognizes as a diftinguifhing mark of its
charaćter. The tongue was often moift, and generally foul; the thirft was feldom great, and though
there was ufually a peculiar naufea, there was rarely
any fevere vomiting or retching. The breathing
was hurried, for the moft part, with much anxiety
and diftrefs; while the patient frequently expreffed
fufferings, which a perfon, unacquainted with the
nature of the difeafe, would be difpofed to believe
were not real.

The fymptoms, which I have enumerated above,
are thofe which ufually fhew themfelves in the firft
twelve hours of the difeafe. I marked them with all
the attention of which I was capable; yet ftill I am
fearful, that the hiftory may not be fo explicit as to be
totally free from ambiguity. The charaćteriftic
marks of the yellow fever, are not by any means
doubtful to a perfon well acquainted with the difeafes
of hot climates; but they are not eafily conveyed in
words, and may often be overlooked by thofe, who have
drawn their information from books alone. I am
induced to think fo, from an inftance which happened
to myfelf. I had read Hillary's account of the yellow fever, both before and foon after my arrival in
Jamaica; I had likewife heard fome converfation on
the fubjećt, fo that I might be fuppofed to have been
tolerably well informed of the general charaćter of
the difeafe; yet the firft perfon, who came under my
care in this illnefs, was within a few hours of death
before I knew the diforder, or even fufpećted it to
be of a dangerous nature. Fortunately for the peace
of my confcience, the patient had been vifited, on
both the firft and fecond day after the attack, by a
praćtitioner who had lived many years in the ifland;
but, between careleffnefs and inexperience, the poor
man's fituation was either not known, or not attended
to, till approaches of death were vifible. The body

had been evacuated very plentifully by a folution of falts, during the two firft days of the illnefs ; but no material good feemed to enfue from it. The patient complained ftill more on the fecond day, than he had done on the firft ; but as the external figns of fever were moderate, I really fufpected that he complained without much caufe. It fo happened, that I could not vifit him on the third ; and on the morning of the fourth, he became of a deep orange colour, and vomited black matter in great quantity. I then fufpected, that this complaint, to which I had not paid particular attention, was actually the difeafe known by the name of yellow fever; but it appeared likewife, to be fo far advanced in its progrefs, that I could do nothing more than witnefs the approach of death. My want of difcernment, and in fome degree my carelefinefs, a charge from which I cannot altogether acquit myfelf, made fo deep an impreffion on my mind, that I turned over every circumftance of the difeafe with which my memory fupplied me; and I foon had that fatisfaction to find, that the miftake which I had committed, had not happened to me without leaving an ufeful leffon. In ten or twelve days another perfon was affected in a manner fo fimilar to the former, that I fufpected the difeafe to be the fame, and the event proved my fufpicions to have been well founded. From that time forward, I never found difficulty in diftinguifhing this particular form of fever, in the firft hours of its attack, not only from the remitting endemic of the country, but even from the other two fpecies of this difeafe, which I fhall afterwards defcribe.

A trifling abatement of the fymptoms, was fometimes taken notice of, in ten or twelve hours after the commencement of this difeafe ; but in no inftance fo far as I have obferved, was there ever fo much alleviation, as with any juftice could be called a remiffion. The fymptoms of diftrefs, where any abate-

ment had been perceptible, recurred in a short time
with aggravation; and if there actually ever was any
relief afterwards, it was only momentary and uncertain. The appearance of the eye became still more
desponding, with a fenfation of burning heat, and
greater marks of inflammation, affording incontefti-
ble figns of the real genius and nature of the fever.
The pain of the head was now violent; the countenance was confufed and grim: the gums were frequently fpongy, and difpofed to bleed; the tongue,
which was fometimes moift, fometimes dry, was almoft always foul; the thirft was irregular; at one
time intenfe, at another very little increafed beyond
what it naturally is. Naufea, I obferved above, was
a common fign, from the firft hours of the illnefs;
yet vomiting, during the firft day or two, was not
by any means a conftant, perhaps fcarcely a frequent
fymptom : neither, if it did take place, was it often
found to be bilious. The liquor thrown up, for the
moft part, was clear; in fhort, feldom altered from
the ftate in which it had been drank, unlefs by having acquired an unufual degree of ropinefs, or by
prefenting fome flakes of a darker coloured mucus.
To the above fymptoms we might add, uncommon
reftleffnefs and anxiety; a torment fcarcely to be expreffed in words; watchfulnefs; a hurried and difficult refpiration; frequent deep and heavy fighing,
with more or lefs difpofition to faint, where any exertion was attempted. It deferves to be remarked,
however, with regard to the difpofition to faint in the
yellow fever, that it did not depend upon the fame
caufe, as in fome other difeafes. It feemed, in fact,
to be owing to torpor of the nervous power, rather
than to excefs of mobility. The patient was often
able to ftand upright, for fome time, even to walk
to a confiderable diftance; and when at laft overcome,
was obferved to fall down in a torpid, rather than in
a fainting ftate. Sweating was a rare occurrence in

this stage of the difease; at least I do not find, that I ever had remarked any greater degree of it, than a clammy moisture on the head and neck. It was also rarely observed, that the external heat was much increased beyond its natural state; while the pulse now began gradually to abate in point of frequency. The yellowness, which is intense in the last stage of the disease, was seldom seen in the period which I now describe; yet, together with a general obfuscation of countenance, a tawney hue rather than a paleness, was observed about the eyes and corners of the mouth, when the patient turned accidentally languid and faint. The body was frequently costive in the first days of the illness; and I have even seen some instances where strong cathartics did not occasion the usual evacuations. The urine was generally high coloured, and turbid. In some cases there was active hœmorrhage from the nose; which was generally followed with some relief from the violent pain of the head. I have also observed a high degree of delirium, though I never saw any instances, where this symptom was of long continuance.

The duration of the tumultuary state, which I have described is uncertain. Sometimes it did not exceed twenty-four hours, though it more generally continued till the third day; sometimes even longer. The symptoms, which now made their appearance, were many of them different in their nature from the former. The agony of distress, which was so strongly depicted in the countenance of the patient, during the first days of the disease, was observed about this time to be sensibly diminished; the eye became more cheerful, the countenance more serene and composed; yet yellowness of the skin became speedily evident: the external heat and fever subsided; the pulse became gradually fuller and slower, and approached by degrees to its natural state: no sweat or moisture was now observable on any part of the body; the state of

Q

the fkin impreffed the idea, as if it were not pervious
to any degree of perfpiration, and heat gradually for-
fook the furface and extremities ; the tongue turned
moift, and at the fame time frequently clean about
the edges : the gums turned redder, more fpongy, and
fhewed a greater difpofition to bleed: vomiting was
now troublefome—the liquor thrown up was ropy,
much in quantity, and abounding with villous or
mucus flakes of a darker colour : thirft in a great
meafure vanifhed ; but fenfations of anxiety, diftrefs
and uneafinefs in the region of the ftomach fuffered
no material abatement.

Things went on in this manner, fometimes for one
day only, fometimes for two, three or more. The
circulation in the extreme veffels became gradually
more languid ; the natural heat retired from the furface of the body, which was now dry and impervious;
the pulfe returned nearly to its ordinary ftate, or
became flow, full, and regular ; the yellownefs increafed faft ; fo that the whole of the body was frequently yellow as an orange, or of as deep a colour
as the fkin of an American favage ; anxiety was in-
expreffible ; vomiting was irreftrainable, and the fo
much dreaded fymptom of vomiting of a matter refembling the grounds of coffee, at laft made its
appearance. It deferves, however, to be remarked,
with regard to this formidable fymptom, that the
colour of what was thrown up, was often black as
foot, where the difeafe had hurried on rapidly to the laft
ftage : while it was not only lefs intenfely black, but
was often tinged with green, where the progrefs had
been flow and gradual. I obferved before, that villous
or mucus flakes were difcovered early in the vomitings of the patient, and that thefe appearances increafed
as the difeafe advanced in its progrefs. I may now
add, that ftreaks of blood were fometimes found to
be joined with them ; the greateft part of which
feemed to come from the throat and gums. The

vomiting, which now returned at fhorter intervals as the difeafe approached this fatal period, was feldom accompanied with violent retching. A quantity of liquor, fometimes a quantity fo enormous, that we could not help wondering whence it had been fupplied, having been collected in the ftomach, was difcharged without much difficulty, and the patient enjoyed fome refpite till a like accumulation had again taken place. It may further be remarked, that as foon as the vomited liquor acquired this dark and footy colour, the belly generally turned loofe, the ftools being black and fmooth, not unlike tar or molafles. The tongue likewife became clean, the gums became putrid; hoemorrhage, or rather oozings of blood were fometimes obferved at different parts of the body; while livid blotches frequently made their appearance on the belly and infides of the thighs. The pulfe, which during the latter ftages of the difeafe, could fcarcely be diftinguifhed from the pulfe of a perfon in health, became at laft quick, irregular, or intermitting; foon after which, coma or convulfions clofed the fcene. It may not be improper to remark, before leaving the fubject, that the yellownefs of the fkin, which was faid to precede the black vomiting in moft inftances, in fome cafes was found to fucceed to it. In fuch, the vomiting began unexpectedly, or without much previous affection of the ftomach; the colour of it was ufually intenfely black, the patient turned yellow almoft in an inftant, and died in a very fhort fpace of time:—the difeafe, in fhort, paffed fuddenly from the firft ftage to the laft.

I may obferve in this place, that the number of thofe who recovered from the laft ftage of this fpecies of the yellow fever, was extremely fmall: yet, though fuch fortunate inftances were rare, they were not altogether wanting. The termination, however, did not appear to be by regular crifis. The black vomiting ceafed, fometimes apparently in confequence

Q 2

of treatment, fometimes evidently of its own accord ;
but a vomiting of a ropy, glutinous matter continued
for a great length of time, together with an extreme
irritability of ftomach, and a very peculiar ftate of
the fkin; which fometimes did not recover its natural fmoothnefs and unctuofity, till after feveral weeks
had elapfed.

During the time that I lived in Jamaica, I opened
feveral perfons who died of this difeafe ; but it was
feldom that I found any material variation in the appearances.  Soon after death, and even fometimes
before death had actually taken place, the body became covered with large livid blotches ; and, it is
almoft unneceffary to mention, was extremely offen-
five.  In opening the abdomen, the omentum and
all its appendages were difcovered to be in a dry and
parched ftate, and of an uncommon dark grey colour.
But together with this dark grey colour of the omentum, and a want of the unctuofity or moifture, which
is ufually found in the cavity of the abdomen, the
ftomach and inteftines had a dirty yellow appearance,
were highly putrified, and much diftended with wind.
The liver and fpleen were generally enlarged in fize;
the colour of the liver was often of a deeper yellow,
than that of any other of the abdominal vifcera ; while
the texture of the fpleen was frequently lefs firm,
than it is found to be in its natural ftate.  The gall-
bladder, for the moft part, was moderately full ; but
the bile it contained, was black and thick, not unlike tar or molaffes.  The biliary ducts were likewife enlarged, and moderately filled with the fame
fort of bile, which was found in the gall-bladder :
while the very blood veffels of the liver bore the marks
of uncommon diftenfion.  In the cavity of the ftomach likewife, there was ufually more or lefs of a
dark coloured liquor, fimilar to what had been thrown
up in the laft ftage of the illnefs.  But befides, that
this dirty fluid was generally prefent in the ftomach

in confiderable quantity, the villous or inner coat of that organ was alfo abraded in various places ; at the fame time that fome fpots appeared on the furface, which were probably the beginnings of mortifications. The fuperior portions of the inteftinal canal were generally in a fituation fimilar to what I have defcribed ; only it muft be remembered, that the morbid appearances were not yet fo far advanced in progrefs.

The ftate of the body, as it appeared on diffection, throws confiderable light on the nature of the yellow fever. It enables us to explain fatisfactorily many of its leading fymptoms; and may even afford ufeful hints in the conduct of the cure. It was mentioned above, that the natural heat and vigour of circulation retired from the furface and extremities of the body at a certain period of the difeafe ; and that a copious and obftinate vomiting enfued foon after this change had taken place. The fluid thrown up, which was ufually pituitous, glutinous, or flakey in the beginning, acquired, after fome time, a colour of various degrees of blacknefs. In quantity, it was often immoderate, bearing no proportion to the liquor which was drank ; a circumftance which could only be explained by the ordina y determination to the furface of the body being turned upon the internal parts ; in confequence of which, there was a preternatural difcharge of fluid into the cavity of the alimentary canal. Flakes, of a mucus or villous nature, were likewife frequently obferved in thofe matters which were thrown up by the patient ; an appearance which we could not have eafily accounted for ; unlefs we had difcovered, in examining the dead body, that the inner coat of the ftomach was actually abraded; but in what manner this happened, may be difficult to explain. It might either arife from the repeated action of fevere vomiting ; or, ftill more probably, from the preternatural and forcible determination to the exhaling veffels of this cavity, forcing

Q 3

off fome portions of the villous coat, in the manner of cuticular blifters. To which explanations I may add, that the black colour of the vomited matter, was evidently owing to a mixture of vitiated bile ; the paffage of which might be eafily traced from the gall duct into the pylorus.

The fpecies of the yellow fever, which I have now defcribed, is univerfally acknowledged to be a terrible difeafe ; and there are few, I believe, fo uncandid, as to boaft of general fuccefs in the manner of curing it. A road is therefore left open, not only for improvement, but almoft for total innovation. The only author I have read on the fubject, or the practitioners with whom I am acquainted, do not feem to have extended their views beyond the fymptoms of the difeafe. There are fome, who, from obferving that there is pain of the head and flufhing of the face, recommend bleeding ; others, from the prefence of naufea or inclination to vomit, make trial of emetics ; and many, from various caufes, infift on the indifpenfible ufe of cathartics. My views, I muft confefs, are different from thofe of preceding authors. Bleeding was employed occafionally ; emetics were cautioufly avoided ; but time appeared to be too precious to be fpent in attending to the effects of cathartics, which cannot often be known in lefs than twenty-four hours ; and which at beft are precarious or feeble. Inftead, therefore, of attempting to evacuate redundant bile, or to correct it when fuppofed to be vitiated, I exerted myfelf, from the firft moment that I was called to the patient, to change the genius and natural tendency of the difeafe ; or, if I may be allowed the expreffion, to take the bufinefs, as fpeedily as poffible, totally out of the hands of nature.

I remark in the firft place, that I generally began the cure of this fpecies of the yellow fever with bleeding. Bleeding was employed in the prefent cafe, chiefly with a view of paving the way to remedies of

greater efficacy. It was, however, found to moderate the violence of local pain, particularly the violence of the head-ach, and to be not altogether without effect, in retarding the ufual rapid progrefs of
the difeafe. It has hitherto been thought neceffary,
indeed almoft indifpenfible, to empty the firft paffages
in this fpecies of fever; but time is fhort, and the
good which accrues from fuch evacuations, is not very
certain, and often not effential. It was, therefore,
thought fufficient to truft this intention, for the moft
part, to laxative glyfters; after the employment of
which, (bleeding having been premifed in fuch quantity as was deemed proper,) the patient was wafhed
clean, and bathed in warm water, in as complete a
manner as the circumftances of fituation would permit. It is needlefs to mention, that this was done
with a view to increafe mobility of fyftem, and to
remove fpafmodic ftricture from the extreme veffels
of the furface; in confequence of which, greater benefit was expected from the application of cold falt-
water, which was dafhed fuddenly from a bucket on
the head and fhoulders. This practice may appear
hazardous, to thofe who argue without experience;
but I can vouch for its general fafety, and bear teftimony to its good effects. Sweat, with perfect relief
from all the feelings of anxiety and diftrefs, was generally the confequence of this mode of treatment.
If employed within the firft twelve hours from the
attack, it feldom failed of removing all the fymptoms
of danger: or of effecting a total and complete change
in the nature and circumftances of the difeafe; but if
the progrefs was more advanced, though the fame
rule of practice might ftill be proper, the execution
required more boldnefs and decifion. It is only poffible to judge from the circumftances of the cafe, at
this period, of the neceffity or propriety of bleeding,
and of emptying the lower inteftines by means of
glyfters; but when this bufinefs fhall have been ac-

complifhed, in fuch manner as may be deemed right, or conducive to the main view, it will be advifeable to fhave the head, to bathe the whole body in warm water, and inftantly to dafh cold water from a bucket on the head and fhoulders. I have even fometimes, where there was an appearance of greater obftinacy, ventured to wrap the whole body in a blanket foaked in fea water, or water in which was difiolved a large portion of falt. If anxiety was great, or naufea and vomiting troublefome, I have alfo obferved benefit from the application of a blifter to the epigaftric region. Opiates, joined with remedies which had a tendency to determine to the furface, were found to be ferviceable; and wine, with a fupply of frefh and cool air, in moft cafes, was highly neceffary. This method of proceeding will, perhaps, be thought unwarrantable; but I can fpeak confidently of its fafety; and I may farther add, that unlefs fome decided fteps are taken to change the nature of the difeafe, during the continuance of this ftage, our future endeavours to do good, will generally be in vain. I have hitherto promifed fuccefs in the cure of this fever, with a good deal of confidence; but if it fhould fo happen, that we are not called to the patient till the yellownefs has fpread over the whole of the body, or till the black vomiting has begun to make its appearance, the profpect, I muft confefs, is then very dark. The ordinary refources of our art are feeble; and if good can be done at all, it can only be done by means, which in the common opinion of practitioners, border on rafhnefs. In this latter ftage of the complaint, fo great a degree of torpor overwhelms the powers of life, that remedies do not produce their ufual effect, and our labour is often the fame, as if we attempted to refufcitate a corpfe. I have, however, feen inftances of fuch unexpected recoveries from the moft hopelefs ftate in fevers, that I feldom totally defpair as long as life remains. I know that death may be prevented, even after black vomiting has appeared with all its terrors, if a remedy can be found powerful enough to excite the action of the extreme veffels, and to recall the determination to the furface of the body. For this purpofe, I have employed alternately warm and cold bathing with fuccefs. I have even wrapt the body, as I mentioned before, in a blanket, foaked in water, in which a large portion of falt was diffolved, or which had been fteeped in brandy or rum, enjoining at the fame time the liberal ufe of wine, or even more powerful cordials. I have heard of fome well-attefted inftances, where plentiful draughts of rum and water, have checked the vomiting, and apparently faved the lives of patients, after the medical people had given them up for loft.

I have now mentioned the method of cure which I purfued in the yellow fever of Jamaica; and I muft be allowed to add, that the general indication appears to be confirmed by a view of the hiftory and progrefs of the difeafe, as alfo by confidering the appearances which are found after death. It was obferved in the preceding pages, that the circulation became languid at a certain period in the courfe of this fever, and that the determination was, in fact, turned upon the internal parts, particularly upon the alimentary canal, and biliary fyftem. To fupport, therefore, or to recall the determination to the furface, where it had begun to retire, was the principal aim which was kept in view. It was purfued with vigour; and, I have the fatisfaction to add, frequently with fuccefs. I am afraid that the means may be thought hazardous; but I have never yet perceived from them, even a momentary harm. I fhall not therefore ceafe to recommend them, till I find that others have tried them fairly, and found them dangerous or ineffectual.

# SECTION II.

IN the preceding pages, I attempted to defcribe the difeafe, which has been ufually regarded as the proper yellow fever of the Weft Indies, detailing at the fame time, the particular fteps of a method of, cure, which I have caufe to believe, was followed with more than ordinary fuccefs. I now proceed to confider another fpecies of diforder, which frequently makes its appearance among people newly arrived in hot countries, and which, from fome ftriking marks of affinity, I have been induced to rank as a fpecies, or variety of the former. Yellownefs, indeed, is not by any means common to it, and black vomiting is actually rare ; yet paroxyfms and remiffions are fcarcely diftinguifhable, and the difference between it and the preceding, is perhaps, in fact, only accidental.

I remarked before, that this fpecies of difeafe, as well as the yellow fever, properly fo called, appears but rarely among thofe who have refided any length of time in tropical climates. It was obferved to begin, as fevers ufually do, with difagreeable affection of the ftomach, with languor, debility, and pain of the head. The horror of fhivering, fo common in the commencement of febrile difeafes, was feldom great in degree ; but it often lafted long, and fometimes was accompanied with unufual feelings. The pulfe was generally fmall, frequent, and eafily compreffed ; the eyes were watery, muddy, or inflamed ; the features were confufed, and the countenance was fometimes flufhed : the thirft was feldom great; and the heat of the fkin was ufually moderate ; but a deep and heavy fighing, a hurried refpiration, with an inconceivable diftrefs and anxiety about the præcordia, gave ftrong indications of the nature of the complaint.—In fome inftances I have known fuch fevere and excruciating fpafms, as, in fome meafure feemed, to fufpend the ordinary functions of life.

In twelve hours, or lefs, there was often fome abatement in the violence of thofe fymptoms; but feldom fuch material relief, as, with any degree of juftice, could be called a remiffion. The fkin became cool, and fometimes moift; yet there fcarcely ever was any fweat. The pulfe became fuller, and often lefs frequent; the reftleffnefs and anxiety were fometimes fenfibly diminifhed; and the local pain often abated; but this refpite was neither long, nor of certain duration. In a few hours, all the fymptoms returned with aggravation. The eyes became more muddy; the countenance more confufed; the headach, and other pains increafed, together with fenfations of anxiety, and reftleffnefs, hurried refpiration, and deep and heavy fighing. The pulfe was now more frequent, fmaller and harder; the thirft was increafed, with naufea, and fometimes with vomiting. The vomiting was feldom bilious: it was not often, indeed, that the matter thrown up, was altered from what had been drank, unlefs by having acquired an additional degree of ropinefs.—To the above fymptoms was fometimes added an obftinate coftivenefs, fometimes fuch a degree of purging and griping, as might eafily be miftaken for proper dyfentery.

As the difeafe advanced in its progrefs, the abatement of the violence of fymptoms, which at firft was fometimes perceived towards the mornings, became gradually lefs and lefs perceptible, and at laft was fcarcely to be diftinguifhed. The anxiety and reftleffnefs were now particularly diftreffing; the fkin was fometimes dry, though oftener moift, and in point of heat below the ordinary temperature of health; while it gave the idea to the perfon who felt it, as if there was a powerful fpafm fubfifting on the furface. I may likewife remark in this place, that a beautiful red colour of the cheeks, together with a fmoothnefs and cherry plumpnefs of the lips, was frequently obferved towards the latter periods of the difeafe. Yellownefs, as was mentioned before, was feldom feen, unlefs in the very laft ftage of the illnefs ; and vomiting of black, or even bilious matter was extremely rare. There was, however, at all times, a great difpofition to faint, with more or lefs of a certain fpecies of low delirium.

The courfe of this fpecies of the difeafe, was lefs rapid, than the courfe of that which terminates in black vomiting ; the termination of the one being often protracted to the eighth or ninth day, that of the other feldom exceeding the fourth or fifth. The marks of crifis, as was obferved before, were rarely difcoverable in the firft fpecies of the yellow fever. They were likewife obfcure in the prefent, and I cannot pretend to fpeak with confidence, of the influence of critical days. Where the termination was favourable, the pulfe became gradually ftronger, and lefs confined in its ftroke ; the fkin likewife became fofter, while the impreffion, which it made on the hand that felt it, communicated an idea that the circulation was more vigorous, and the fpafm on the furface lefs obftinate; the eye and countenance likewife brightened up; the anxiety and reftleffnefs vanifhed or decreafed, and fome appetite for food returned ; but it was often difficult to mark the point of time precifely, at which this change took place.— It may be obferved likewife, where the termination was fatal, that death approached in two different ways. A patient, apparently poffeffed of vigour, was fometimes fuddenly feized with coma or convulfions, and died unexpectedly ; but it happened more frequently, that the powers of life were gradually and flowly extinguifhed; the pulfe became weaker and more confined in its ftroke; while the natural heat and circulation retired by degrees from the furface and extremities of the body.

The cure of this fpecies of the difeafe, though by no means eafy, was lefs difficult upon the whole than that of the former. Inftead of the torpor and infenfibility, which prevailed in the latter periods of the proper yellow fever, the mobility of the nervous fyftem was fo much increafed in the prefent fpecies of difeafe, that remedies feldom failed of producing fenfible effects : and wherever remedies produce effects, it generally is in our power to manage the bufinefs in fuch manner, that fome good may arife. It may be obferved in the firft place, with regard to the cure, that bleeding, which frequently was ufeful in the former fpecies, was generally hurtful in the prefent ; and that inftead of retarding, it oftener accelerated the progrefs of the difeafe. Emetics were employed very commonly by the practitioners of the Weft Indies, in this as in other cafes of fever ; but I cannot help remarking, that languor and debility, frequently yellownefs, and fometimes a continual vomiting, which no remedies could reftrain, were often the confequence of antimonial emetics of fevere operation ; and I have no doubt in faying, that the approach of death was actually haftened, in feveral inftances, by this method of treatment. Laxatives were occafionally of fervice ; but the ftronger purgatives were frequently hurtful. Blifters were often extremely beneficial ; but it requires care and difcernment to apply them in the proper circumftances, fo as to reap the full advantage. Opiates were fometimes ferviceable, and bark and wine, in moft inftances, were remedies of great value ; but the principal truft was placed in warm and cold bathing ; which, under proper management, feldom failed of anfwering every expectation completely, or fpeedily, of removing the chief fymptoms of danger. Sometimes it appeared to cut fhort the courfe of the difeafe abruptly.

R

# SECTION III.

I HAVE now defcribed two fpecies of a fever, which feems to be, in fome degree, peculiar to the natives of northern regions, foon after their arrival in the Weft-Indies. In the one, a determination to the alimentary canal and biliary organs, with marks of putrefcent tendency in the general mafs of fluids, was difcoverable at an early period ; in the other, the brain and nervous fyftem were more particularly and principally affected; while the fpecies, of which I now attempt to give fome account, exhibited ftrong marks of vafcular excitement, with a very high degree of the apparent inflammatory diathefis. This was more irregular in it. appearances and more complicated in its nature, than the others. The marks of inflammatory diathefis were generally very apparent in the beginning ; but they ufually gave way or became complicated in the latter ftages, with fymptoms of putrefcency or nervous affection. In defcribing the hiftory of this difeafe, it may not be fuperfluous to remark, that there is feldom any thing particular in the fenfations of debility and horror, which precede the formation of the paroxyfm. The hot fit was generally obferved to run high ; the heat was often intenfe ; the pulfe, which was quick, frequent and irregular, vibrated often in an uncommon manner, and with an ufual degree of force; the thirft was fometimes immoderate, fometimes not greatly increafed ; the countenance was flufhed ; the eye gliftened, and appeared frequently to be in fome degree inflamed ; the figns of excitement were in general uncommonly high ; yet the difpofition to faint was fometimes fudden and unexpected. It deferves farther to be remarked, that blood drawn from the arm did not commonly exhibit the ufual buffy appearance of real inflammatory diathefis; and though times of aggravation and alleviation were often difcernible; yet they did not happen at regular and ftated periods.

It was obferved frequently, that many of the leading circumftances fuffered a material change, about the third day of the difeafe. The fymptoms of high inflammatory diathefis, which prevailed in the beginning, became mixed, more or lefs, with fymptoms of putrefcency, or nervous affection. Delirium made its appearance; fometimes it ran high, with ftartings and fymptoms of violent excitement; fometimes there was a low and muttering incoherence with marks of languor and debility. The gums turned red and fpungy, and fometimes bled; the thirft was frequently intenfe, the tongue dry, with vomiting and fevere retching; yet vomiting of bilious or vitiated matters was a rare occurrence. The above fymptoms generally went on to increafe, during the fpace of fix or feven days, about which period the powers of life either yielded to the difeafe, or figns of recovery began to appear: the marks of crifis, however, were feldom diftinct and final; neither was the influence of critical days fo much to be depended upon as in the common remitting fever of the country.

It was mentioned above, that the nature of this fever was more complicated than that of the two former; fo the indications of cure are likewife more difficult and perplexed. If we proceed on the firft obvious view of the difeafe, we fhall often do irreparable mifchief by copious and repeated evacuations; yet there will not be lefs danger, on the other hand, if, regardlefs of the prefent degree of excitement, we indulge freely in the ufe of ftimulants. It is neceffary to obferve a middle courfe; and I muft confefs, that it is fometimes difficult to do any thing, without doing harm. Bleeding was frequently employed in the cure of this difeafe, and in moft cafes, it was a ufeful remedy, though lefs perhaps from its own effects merely, than from paving the way to other more powerful applications. It is, however, capable of being eafily carried to excefs; and ought not to be trufted to wholiy for the removing of the irritability, and high degree of excitement, which prevails fo generally in the beginning of this difcafe. After bleeding, emetics and cathartics are employed ver, freely. I have always profeffed myfelf an enemy to the practice of giving emetics in the fevers of Jamaica; yet, I muft confefs, that antimonials were not only fafer, but of more particular fervice in this, than in any other fpecies of fever, where I have feen them tried. Among the great variety of forms which have been recommended by practitioners, for the purpofe of emptying the firft paffages, I have not found any one anfwer fo well, as a thin folution of the fal-catharticum, given at different intervals, with a fmall portion of emetic tartar, and fometimes with the addition of laudanum. The operation of this remedy was extenfive. It might be fo managed, as to promote naufea or vomiting, fweat, or moderate evacuations downwards; at the fame time that it proved very powerfully fedative. I may likewife add, that I have fometimes found benefit from nitre, camphire and opium, given in pretty large dofes, and accompanied with plentiful dilution. But though thefe remedies were often ferviceable, and contributed in many cafes to moderate the high degree of irritability; yet the chief dependence of the cure was much better trufted to cold bathing. After the furface of the body had been fufficiently relaxed, by the previous ufe of warm bathing and fomentations, the effects of cold bathing were wonderful. The exceffive irritability was moderated or removed, and the powers of life were invigorated in a very fingular manner in confequence of it.

I have attempted in the preceding pages, to give a fhort view of a difeafe, which has not, I believe, been hitherto very accurately defcribed by authors' or treated with much fuccefs in practice. It is a difeafe of a continued kind; and, as I faid before, in fome degree peculiar to the natives of northern latitudes, foon after their arrival in the tropical climates. I cannot help thinking, that it may be eafily diftinguifhed, even in the firft hours of its attack, from the intermitting or remitting fever, which is the common endemic of hot countries: but I muft at the fame time add, that this diftinction does not refide in the prefence or abfence of one individual fymptom.. The ftate of the pulfe, indeed, conveys information,. that the difeafe is not of the kind which has paroxyfms and remiffions : yet this information can only be obtained from a knowledge and actual comparifon of the two difeafes :—I do not pretend to defcribe it in words. The ftate of the eye and countenance, was likewife obferved to be ftrongly defcriptive of the na ure of the difeafe; as alfo were the deep and heavy fighing, the hurried refpiration, the anxiety and reftleffnefs, with a certain uncomfortablenefs of fenfation, which no words can exprefs ; but I confefs myfelf, at the fame time, perfectly at a lofs to fix on any one fingle fymptom, which appearing at an early period, difcriminated it with certainty from all other fevers. I have defcribed it under three diftinct and feparate forms; but I muft alfo add, that they may fometimes be found to be more complicated with each other, than they appear to be in the above defcription..

R 3

# C H A P. XII.

HAVING endeavoured in the preceding treatife, to give a more accurate hiftory of the endemic fever of Jamaica, than is met with in books, and I am difpofed to flatter myfelf, having pointed out a more fuccefsful method of cure than that which has been generally purfued; I fhall now add a few obfervations on the intermitting fever of America; a difeafe, in which my experience has been tolerably extenfive. The frequent occurrence of intermitting fevers in every climate, together with the full and ample manner in which the difeafe has been treated of by many learned and ingenious wri ers, excufes me from entering into a minute and full difcuffion of the fubject. I fhall therefore employ only a few pages in attempting to illuftrate particulars in the hifto y of the difeafe, which have been fuperficially noticed; or to explain fome points of treatment, which, though not new, I have ventured to carry farther than is ufual in common practice. As I had the opportunity of attending to the hiftory of the intermitting fever in feveral of the fouthern provinces of the continent of North America, I fhall firft mention the more conftant and general courfe of the difeafe, and afterwards point out thofe circumftances of peculiarity, which feemed to arife from the difference of climate, or from the influence of the feafon of the year. I fhall likewife occafionally take notice of the general ftate of health of the troops who were employed on the fame expedition,

though I muft alfo add, that I can only pretend to
trace the progrefs of the fever with accuracy, in the
regiment in which I had the honour to ferve.

I fhall attempt, in the firft place, to give an accurate
defcription of the paroxyfm of an intermitting fever,
marking as carefully as I can, the order of fucceffion,
in which the fymptoms moft ufually appear. We
are taught by the defcriptions of moft writers, to confider languor and debility as the firft feeling or firft
effential fymptom in the paroxyfm of an intermitting
fever; but I cannot avoid remarking, that an unufual
affection at ftomach, a flatulence,—in fhort, fomething
difagreeable, which I cannot eafily define, but which
was accompanied in many cafes with head-ach, and
fometimes with drowzinefs, preceded every fenfation
of languor or debility in moft cafes, where my obfervations were made with fuch care that they could be
trufted to.    I may alfo farther obferve, that, as foon
as this languor or debility began to be perceived, the
veins began to fubfide, the nails turned pale, and at
laft blue; the fkin of courfe was dry and conftricted;
and there was fometimes an evident diminution of
heat, particularly of the heat of the extremities.    To
thefe fymptoms was often added, a difagreeable kind
of yawning, with ftrong fenfations of wearinefs, and
an irrefiftible inclination to ftretch the limbs.    A
fenfation of cold was now felt in the back, as if water
ran down upon it in feparate ftreams.    It foon vanifhed, indeed; but fuddenly returned again in a more
violent degree; in which manner it went on, ceafing
for an inftant, and then recurring with aggravated violence, till the whole body became at laft affected with
rigour or fhaking, accompanied, in a more efpecial
manner, with chattering of the teeth.    The coldnefs
having now arrived at its acme, or higheft point of
intenfity, glowings of heat were perceived in the intervals between the rigors or fucceffions. Thefe glowings
grew gradually ftronger, and continuing for a greater

length of time, at leaſt baniſhed every ſenſation of
cold. The heat, which now ſucceeded, was often
much above the temperature of health; marks of fever ſometimes ran high; the veins became full; the
face was fluſhed, and the ſurface of the body bore
marks of diſtenſion. The duration of this ſtate was
uncertain: ſometimes it did not continue the ſpace of
one hour, ſometimes it laſted four or five. A damp-
neſs at firſt began to appear on the forehead and
breaſt, which extending itſelf gradually to the extremities, was at leaſt formed into a ſweat; in conſequence
of which, the fever gradually ſubſided, and the body
returned nearly to its natural ſtate.

The above are the moſt uſual ſymptoms of the paroxyſms of an intermitting fever. I have deſcribed them
in the order of time in which they moſt uſually appear. I muſt however remark, that ſymptoms are
ſometimes obſerved different from thoſe which I have
now taken notice of; as alſo, that the order of ſuc-
ceſſion, which I have mentioned, is not, by any means
fixed and invariable. It is impoſſible to deny the
common obſervation, that languor or debility is a general and early ſymptom in almoſt every ſpecies of
fever; but it is likewiſe certain, that there are many
inſtances, where it is not in our power to perceive its
actual preſence. It is therefore precipitate to conclude with Dr. Cullen, that all the future phænomena depend upon this, as their eſſential and original
cauſe. There may frequently be deception in attempts to deſcribe the ſituation of others; but that
which we feel ourſelves is more to be truſted to:—
and I can affirm, that I have often felt ſenſations of
cold in my own perſon, previous to every feeling of
languor or debility; previous, I might even ſometimes ſay, to any perceptible deviation from a ſtate of
health. But beſides, that the exiſtence or perception
of languor and debility, does not ſeem to be eſſential to the exiſtence of a paroxyſm of intermitting

fever, I may likewife add, that I have feen inftances, particularly in the hot.months of fummer, where the whole of this difeafe paffed over, without the leaft perceptible degree of a previous cold fit. It happened fometimes alfo, that during the continuance of the paroxyfm, there was fcarcely any obfervable diforder in the pulfe, or any material figns of external fever. The tumult and uneafinefs, which terminate in moft cafes by fweat, went off in fome by urine or ftool, or perhaps declined in others, without the appearance of any preternatural evacuation. In like manner it was commonly obferved, in the difeafe diftinguifhed by the name of partial intermittent, that there was not any perception of cold, nor increafe of heat; no diforder in the pulfe, or preternatural evacuation; in fhort, not a fymptom, which characterizes the genius of the difeafe, except local pain, which continuing for a certain time, difappears, and then returns again at a ftated hour. To this we may add, that there are various inftances, where the whole duration of a complaint, which indifputably depends on the caufe of intermitting fever, is occupied by a comatofe difpofition, by convulfions, or even by tetanic affection. If we therefore confider thefe phænomena attentively, we fhall find little caufe to believe, that the moft ufual fymptoms of the intermitting fever, are fymptoms without which the difeafe cannot exift; or that they are mutually the caufe and effect of each other. The order of fucceffion I have obferved is not fixed invariably; and cafes are numerous, where thofe fymptoms, which fome authors have confidered as abfolutely effential, do not appear at all. This fact is certain; and we may fafely conclude from it, that the main hinge of action in a paroxyfm of fever has not been yet difcovered.

The vital and natural functions are varioufly affected, not only in different people, but in the fame perfon, in the different ftages of a paroxyfm of the

fame fever. The pulfe, in the firft approach, is often remarked to be flower than natural, fometimes it
is more languid and weak. It foon however becomes
more frequent, though it continues for the moft part
fmall and contracted, till the latter ftage of the cold
fit. It then ufually acquires ftrength and fome degree of fulnefs, fometimes greater frequency and
hardinefs; but as the fweat begins to flow, the hardnefs and frequency abate, while the fulnefs increafes;
fo that it returns by degrees nearly to its natural ftate.
The difagreeable affection of ftomach, (which I formerly obferved was fometimes the firft perceptible
fymptom of a paroxyfm of the intermitting fever) increafes frequently to naufea or retching, fometimes
to fevere and continual vomiting; which does not
ceafe till fweating has become general all over the
body. The refpiration, which in the beginning of
the paroxyfm, is ufually flow, and fometimes interrupted with fighing, in the progrefs of the hot fit
becomes frequent, laborious and high. It often happens, likewife, that there is more than ordinary dul-
nefs of perception in the mental faculties in the firft
approach of the fever; while this is often fucceeded
by extraordinary acutenefs in the more advanced
ftages, particularly during the continuance of the hot
fit. But though it is only during this period that
excitement and delirium are obferved to be common;
yet inftances are not wanting, where derangement of
intellect is among the firft fymptoms of the difeafe,
and where it continues among the principal throughout the whole of the courfe. To the above appearances we may add, that the urine, which is thin and
pale in the firft ftage, becomes high coloured in the
progrefs of the hot fit; and as the fweating advances,
thick and turbid, frequently with the addition of a
copious lateritious fediment.

I remarked formerly, in treating of the remitting
fever of Jamaica, that certain hours of attack were

in a very peculiar manner connected with the different types or forms of that difeafe; but I cannot pretend to fay, that the fame rules were obferved to hold good, with any degree of certainty in the intermitting fever of America. Single tertians, indeed, began moft ufually about twelve; though there were likewife many inftances where they came on fo early as ten in the morning, or fo late as two in the afternoon. The other forms were ftill lefs regular. It was alfo taken notice of, that anticipations were common in the fingle tertian of Jamaica; as alfo that they were irregular and long. In America they were ftill more frequent; but feldom exceeded one or two hours at once. They often, however, prevailed to a certain acme, or point in the difeafe, obferving a regular interval of time in their progrefs. It fometimes likewife happened, that the type poftponed gradually, till the complaint difappeared finally, This, however, was much more rare than the other.

Having mentioned, in the preceding pages, fome general refemblances of the intermitting fever of America, I fhall next trace its peculiarities in the different provinces, in which the regiment to which I belonged, had the fortune to ferve. I may obferve, in the firft place, that I joined the firft battalion of the 71ft regiment, on York Ifland, in the beginning of the fummer 1778. Few of the men were then fick; neither did the number increafe materially, till towards the latter end of June. The intermittents, which appeared previous to this period, were generally fingle tertian; and of perfectly eafy treatment. In the month of July, a dyfentery, of a very particular kind, became epidemic, and the fporadic intermittent inftantly vanifhed. The ftools in this complaint were numerous and bloody, the gripings were fevere, but there was feldom any very material diforder in the pulfe. The difeafe did not often terminate in lefs than feven days; fometimes it continued a

fortnight or longer. The ordinary treatment was
very rarely of benefit; yet the complaint was of a
nature fo little dangerous, that I do not recollect a
fingle perfon who died of it. It difappeared totally
about the beginning of Auguft, or rather changed
into an epidemic intermittent, the type of which was
ufually fingle tertian. The paroxyfms of this fever
were regular, the intermiffions were diftinct; and its
nature was fo far from being obftinate, that I fcarcely
met with an inftance which refifted the Peruvian
bark, where that remedy was given in fufficient
quantity. This fever continued highly epidemic
during the months of Auguft and September. The
frequency of new attacks was confiderably diminifhed
in the month of October; yet fuch as happened then,
were generally accompanied with dangerous and
alarming fymptoms. Relapfes were common. But
though the intermitting fever of this ifland was epidemic in a confiderable degree, it was not by any
means of a fatal nature. If neglected in the beginning, foundation was fometimes laid for obftinate
complaints; but the difeafe was not fatal in its proper
form to any one patient, who remained with the
regiment. I cannot fpeak with certainty of the iffue
of a few of the worft cafes, which were fent to the
General Hofpital, on the breaking up of the encampment in the month of November. The regiment
was then embarked in tranfports, on an expedition to
the fouthward. The fick were collected into one
fhip, which, after a ftormy and tedious paffage, arrived
with the reft of the fleet at Savanna, in Georgia, in
the latter end of December. The voyage had an
excellent effect on the health of the men. Out of a
hundred and twenty convalefcents, who embarked at
New York, in the month of November, not a man
died; and there only remained two, who were unfit
for the fervice of the field, on the day of our arrival
in the Savanna river. During the months of January,

February and March, the battalion of the regiment
in which I ferved, was a total ftranger to ficknefs.
It was employed in long and almoft continual marching, till the latter end of April, when, encamping at
Ebenezer, on the Savanna river, the intermitting
fever foon made its appearance, and fpread fo rapidly,
that before the end of June, very few remained, not
only in this regiment, but even in the garrifon, who
had not fuffered more or lefs from this raging difeafe.
It was commonly remarked in the hiftory of this
fever, that the type during the month of May, was
ufually fingle tertian, till the fifth or fixth day; after
which, paroxyfms were often obferved daily, though
generally unequal in force and duration : that is, the
difeafe changed about this period, to a double tertian
form. But though this was obferved to be the cafe,
during the greateft part of May, the type of the fever
was ufually double tertian, or quotidian, from its
very commencement, in the month of June. The
difeafe was then of the moft ardent kind. The paroxyfms were feldom ufhered in by a cold fit; and
the remiffions, for the moft part, were very indiftinct
and imperfect. The heat of the weather was exceffive, during the greateft part of the month; and
ftrange and alarming fymptoms occurred frequently
in the courfe of the difeafe. In fome cafes a comatofe
difpofition, approaching to apoplexy, or rigid fpafms,
refembling a perfect tetanus, occupied the greateft
part of the paroxyfm; in others there were various
local pains, deliria, bilious vomitings or purgings,
with a multitude of other affections, which appeared
on a fuperficial view to conftitute the whole of the
complaint. Yet thefe fymptoms declining after fome
continuance, recurred again at a ftated hour, and
were finally removed, or at leaft fufpended, by the
Peruvian bark. I left the garrifon of Ebenezer in
the beginning of July, and went directly to Savanna,
where the fame epidemic prevailed, though in a de- -

S

gree of lefs frequency, and with fymptoms of a lefs alarming nature than at the above-mentioned place. At Savanna, it ufually retained marks of diftinct intermiffion, and its type was often of the fingle tertian kind—in fhort, it was fimilar to the fever of Ebenezer in the month of May. From Savanna, I went to Beaufort in the beginning of Auguft. The fever, which ufually prevails at this feafon of the year, in all the fouthern provinces of North America, was then epidemic among the troops who were ftationed on this ifland. The type, however, was ftill more commonly fingle tertian here, than at Savanna. The beginning of the paroxyfms was likewife more generally diftinguifhed by a cold fit; and the intermiffions, for the moft part, were more perfect and diftinct. In a few cafes, indeed, marks of malignity were difcoverable; yet the difeafe, upon the whole, was not of a fatal nature, or of obftinate cure; though unlefs fpeedily checked by bark, it often degenerated into dyfentery or dropfy, which were not only removed with difficulty, but in the circumftances under which we laboured, were often of very precarious iffue. This epidemic was ftill acquiring force, when the outpofts were fummoned to the defence of Savanna. Its progrefs was, in fome meafure, fufpended during the active fervice of the fiege. The enemy, however, had no fooner retired from before the place, than a fever began to rage with violence, which carried off prodigious numbers, particularly of the foreign troops. It was obferved in the hiftory of the preceding year, that few were attacked afrefh with the intermitting fever on York Ifland, fo late as the months of October and November; but it was likewife remarked, that, where the difeafe happened at thofe periods, the fymptoms were oftener malignant or dangerous. The fame was in fome refpects the cafe at Savanna. The fever, which made its appearance after the fiege, was of an alarming and violent kind. Marks of diftinct intermiffion were feldom difcoverable, delirium was a common fymptom, fpafmodic affectioas were fome-times violent, and the courfe of the diforder was ge-nerally rapid. The rage of this epidemic ceafed in December; but relapfes continued to return occa-fionally, during the following winter; which was an unufually fevere one in that fouthern latitude.

There likewife ftill remained fome dyfenteric com-plaints, which refifted every mode of treatment that could be devifed. They yielded however to the re-turn of the warm weather, affifted, in no fmall degree perhaps, by the active fervice of the fiege of Charlef-town. The recovery, indeed, was fo complete, that, in the beginning of June, the whole of the regiment arrived at Camden in perfect health. The firft bat-talion was fent to occupy a poft at the Cheraws, on the river Pedee. The diftance is feventy-five miles; yet fuch was the fpirit and activity of the men, that they performed the march in three days, without fatigue or inconvenience. An open field, between four and five hundred paces from the bank of the river, was chofen for the encampment of this battalion; while a fituation perfectly dry and cleared of wood, but nearer to the bank, was referved for the encamp-ment of the fecond, which was not expected to ar-rive till after fome time. In a fortnight or three weeks, the intermitting fever began to fhew itfelf. It fpread fo rapidly, particularly in the fecond bat-talion, that before the end of July, when the poft was abandoned, few were left who had not felt its influence. The prevailing fymptoms of this difeafe were much fimilar to thofe of the fever of Ebenezer. The type was frequently double tertian, or quoti-dian; the remiffions were indiftinct; the bilious vo-mitings and purgings were often exceffive, and marks of malignity appeared in feveral inftances. The approach of the enemy made it neceffary that the poft fhould be withdrawn; but there was much

difficulty in accomplifhing it. Two thirds of both
officers and men were unable to march; and it was
not poffible, in the fituation in which we were placed,
to find waggons fufficient to carry them, together
with the neceffary provifions and baggage; fo that
no other refource was left, than to convey fome part
of them to George Town by water. Boats were
therefore collected for this purpofe, and fuch men
were put into them, as were judged leaft likely to be
foon fit for the fervice of the field. Thefe, however,
unfortunately fell into the hands of the militia, in
their paffage down the river, and were foon difperfed
into the different parts of the country; fo that I cannot fpeak with certainty of the general iffue of the
difeafe. Thofe who retired to Camden by land, improved unexpectedly in the ftate of their health, in
the courfe of the march. During the time that we
lay at the Cheraws, the remiffions were generally
obfcure; but after the fecond or third day's march,
the type changed frequently from double to fingle tertian; at the fame time that intermiffions became clear
and diftinct. It may be difficult to determine precifely to what caufe this might be owing; whether
to removal from a fituation, where the fomes of the
difeafe was in a very concentrated ftate; to the mere
exercife of travelling; or to the effects of cooler
weather with rain, which happened at this time, and
which continued for two or three days with little intermiffion. The whole of thofe caufes, perhaps,
contributed to operate this falutary change; though
it will probably be reckoned among the firft inftances,
where travelling and getting wet, are recommended
as being ufeful in the cure of fevers. During the
month of Auguft, and a great part of September,
the army remained encamped near Camden. The
weather was excefiively hot, and fevers were frequent,—fometimes malignant and dangerous; though
they preferved, in general, the diftinct character o

intermittents. In the months of October and November relapfes were numerous, and original attacks,
though rare, were dangerous and alarming when they
happened. Some inftances of a difeafe were now
obferved of a more ferious nature than any that had
hitherto appeared. Inftead of diftinct intermiffions,
which prevailed during the preceding months, the
fmalleft traces of remiffion were fcarcely perceptible;
the countenance was dufky, and of a greafy appearance, the tongue was conftantly dry and parched;
the head was often much affected, and grangrenous
fpots fometimes appeared on the extremities. The
duration of this difeafe often did not exceed feven
days; fometimes it continued a fortnight, or even
longer. It was generally of a fatal nature; and
where it happened to people who had been fubject
to the intermitting fever in the preceding months,
it for the moft part effected fuch a change on the
conftitution, as deftroyed the tendency to relapfe.
But befides this unufual fpecies of difeafe, which
fometimes appeared in the months of October and
November, it was likewife obferved that relapfes of
the fever, which preferved the diftinct intermitting
character, were not only lefs frequent, but commonly
lefs alarming, in proportion as the weather turned
cooler. Relapfes were often remarked in this feafon to terminate of their own accord, in a very fhort
time; and frequently to leave the body in a ftate of
greater vigour than they found it. I find a fact in
my notes, with regard to this fubject, which is curious
and important. Between thirty and forty of the
men of the regiment entered upon the fervice of the
campaign in fo weak a ftate, that they were unable
at firft to carry their arms. They however gained
ftrength fpeedily as they proceeded on the march;
and feldom forgot to mention, that they felt a new
acceffion of vigour after every accidental relapfe.
But I muft further obferve, that, together with the

above changes which happened in the progrefs of the feafon, the epidemic fhewed a remarkable tendency to degenerate into dyfentery or dropfy in the months of September and October. The gripings in this fpecies of dyfentery were often fevere; the ftools were large and watery; and times of aggravation and remiffion were frequently obferved, as in a regular intermittent. Indeed the intermittent, the dyfentery, and even the dropfical fwellings fo often alternated with one another, as evidently fhewed that they all depended upon the fame general caufe. The campaign of the following winter was a very active one. The army travelled over a great extent of country, and was confidered by many as performing very hard fervice; but I have the fatisfaction to add, that notwithftanding occafional forced marches, wading of rivers, expofure to rain, accidental fcarcity of bread, and no great profufion of beef, with the total want of rum, the troops enjoyed in general a moft perfect ftate of health. Valetudinarians were reftored to perfect vigour; and when we arrived at Wilmington, in the latter end of April, there fcarcely was a man in the regiment to which I belonged, who was not fit for the duty of the field. In the fummer campaign through North Carolina and Virginia, there was no room to complain of hardfhips. The camp abounded with a profufion of the beft provifions; and the marches were feldom long or fatiguing. We arrived at Portfmouth towards the end of July, with a very moderate lift of fick. Portfmouth is faid to be unhealthy; and we foon were able to verify the obfervation: an intermitting fever, complicated, or alternating with a dyfenteric complaint, made its appearance foon after our arrival, and continued to increafe during the fhort time we remained in the place. A difeafe of a fimilar kind continued to prevail in the army, after our removal to York Town. It was not, however, by any means fatal in its nature, or difficult of cure, if attended to in time, though if allowed to go on, it often degenerated into dropfy, obftructions in the abdominal vifcera, or a dyfenteric complaint which frequently proved fatal in the beginning of the following winter. The 71ft regiment had now ferved three campaigns in the fouthern provinces, and might be confidered as being perfectly well feafon d to the climate. It was in fact more healthy than any other corps in the army; there not being more than five or fix unfit for the duty of the line, when the French and Ame icans invefted the place. After the capitulation the proportion of the fick of the army increafed confiderably. Some inftances of a fever, fimilar to that which prevailed at Camden and Savanna, in the month of November, were obferved in feveral regiments; but a fpecies of dyfentery, which appeared often to have originated from an ill cured intermittent, was the complaint which proved principally fatal.

From the above fhort hiftory of the intermitting fever, as it appeared in the 71ft regiment, in the different provinces of North America, where that corps happened to ferve, we may be enabled to form fome idea of the changes, which are more conftantly produced by feafon and climate, or which arife accidentally from the particular effects of local fituation. In the fpring and beginning of fummer, the fingle tertian was the moft ufual type of the endemic of America, in every province which the regiment vifited:— the paroxyfms were diftinct, and the intermiffions were generally perfect. In the months of June, July and Auguft, double tertians were common, and in fome fituations banifhed every fimpler form. As the weather turned cool, the fingle tertian refumed its place; fo that any other type was fcarcely ever feen. But befides the above changes of type, which in fome degree followed the changes of feafon, dyfentery or dropfy frequently made their appearance in the months of Auguſt, September and October, alternating with, or ſucceeding the intermittent; while fevers of a bad and uncommon kind were by no means rare in the months of October and November.

I have thus obſerved in a curſory manner the more general changes of the intermitting fever, as influenced by change of ſeaſon. I may alſo remark, that beſides ſeaſon, climate had a conſiderable effect in modifying the appearances of the diſeaſe. It thus happened, that the type was generally ſingle tertian on York Iſland, even in the heat of ſummer; in ſpring and winter other forms were rarely ſeen. In Georgia, the ſingle tertian was the prevailing form, only in the winter and ſpring. In ſummer, and ſome part of autumn, double tertians were common; and types of ſtill greater complication frequently made their appearance during this period, in ſome particular ſituations of the province. Dyſentery dropſy and dangerous fevers were likewiſe more frequent here in the autumnal months, than they were found to be in the neighbourhood of New York; while the courſe of intermittent, as long as the form was regular, was more ſpeedily checked by Peruvian Bark in Georgia, than in the more northern latitudes. The prevailing type of the climate of South Carolina, was ſingle tertian, even in the ſummer and autumn; yet where the forms of the diſeaſe was in a high ſtate of concentration, as at the Cheraws, the type was often ſo complicated that remiſſions were ſcarcely diſcernible. The tendency of the endemic of this province, to degenerate into dyſentery or dropſy in the autumn, was likewiſe leſs remarkable than in Georgia. The dangerous fevers of October and November were alſo fewer in number; though ſtill more frequent, and more formidable than in the province of New York. The province of Virginia lies about halfway between New York and Savanna; and the general effects of its climate, on the common endemic of the country, correfponded with its local fituation.—Deviations from the tertian type were more frequent than at the one place, lefs fo than at the other.

It appears from what has been faid above, that the fingle tertian is the proper fundamental type of North America. It undergoes, as we have feen, a regular change and alteration, in confequence of the ordinary changes of the feafons, as alfo in confequence of the effects of the various climates of the different provinces of that extenfive continent: but befides thefe changes, which are more general and certain, we likewife find, that the accidental circumftances of local fituation often produce very remarkable effects. In this manner, though the type of the fever which prevailed on York Ifland, was properly fingle tertian; yet double tertians, and even more complicated forms, were not by any means rare, in a part of the battalion which lay contiguous to a fwamp. The real nature of the endemic fever of Georgia, is, perhaps, properly of the intermitting kind; yet remiffions were often fcarcely perceptible at Ebenezer; which is fituated immediately on the bank of the river Savanna, and which, in fome degree, is furrounded by creeks of frefh water. It may not be improper to remark with regard to Ebenezer, that few places in America have been obferved to be more unheaithful; though fuch a conclufion probably would not be drawn from a general view of its fituation. It occupies a fandy eminence of confiderable elevation, and poffeffes a confiderable environ of cleared ground. At Savanna, which is fituated twenty-five miles nearer the mouth of the river, there were likewife many inftances of deviation from the fingle tertian type, but ftill fewer than at Ebenezer. The fever likewife was generally of a lefs dangerous kind. The obvious appearances of the two places did not afford fufficient reafon for forming this conclufion. The fituation of Savanna would have probably been thought

to be the leaft favourable to health. Though elevated
and dry, and poffeffing a wider environ of cleared
ground than Ebenezer; yet a fwamp on the right and
left, with a river and rice fwamps in front, threatened great ravage from intermittents. That they
were lefs formidable than might have been expected,
was probably in a greater meafure owing to the bluff
or fand bank being higher than the fituation of the
town, and intercepting, in fome degree, the exhalations from the river and great fwamps.

I obferved on a former occafion, that the figns
of crifis, in the remitting fever of Jamaica, were
generally clear and unequivocal. I muft now own,
that I have not been able to attain certainty, on this
head, in the intermitting fever of America. In fevers of a fingle tertian type, the intermiffions were
frequently fo perfect and complete, that it was not
eafy to fay what was wanting to conftitute perfect
health: even in the hot months of fummer, where the
remiffions were extremely obfcure, I often found it
difficult to form an opinion to which I could confidently truft ; as it happened frequently, that thofe
figns, which I had been difpofed to confider at one
time as marks of final crifis, proved in the event only
to be indications of more diftinct intermiffion, or of
fome change in the nature of the fymptoms.

Having given a fhort view of the hiftory and progrefs of the intermitting fever of America in the preceding pages, I fhall now proceed to offer a few obfervations on the manner of treatment. And I may
obferve in the firft place, that the intermitting fever
is not in general a difeafe of a dangerous nature. If
treated with decifion in the beginning, it is for the
moft part removed very fpeedily and very certainly ;
though if attacked with feeble remedies, it often continues long, and not feldom lays the foundation of
complaints which eventually have an unfavourable
termination. The intermitting fever fometimes proves

fatal from the actual violence of the fymptoms of the
paroxyfm, though the danger more generally arifes
from a tendency to degenerate into dyfentery or dropfy, or to form vifceral obftructions. But befides the
danger, which arifes from the actual force, or from the
more tedious effects of the difeafe, we-often find a
character of peculiar malignity, in the intermittents
of fome feafons and fome fituations, which deferves
to be particularly attended to. Malignity is a word of
a vague meaning; and on different occafions is differently applied. In the prefent inftance, I refer the
term to a peculiar character of the difeafe, exprefled
by a certain ftate of the eye and countenance of the
the patient. It was fometimes obferved, that the
countenance of the patient was flufhed; but at the
fame time dark and overcaft; or that it was of a greafy
and dufky appearance, with a look of fternnefs and
defpondence in the eye. Thofe appearances, particularly where a white glutinous covering appeared on
the tongue, were conftantly indications of concealed
or lurking danger. They occurred often at Ebenezer and the Cheraws, and I had the misfortune to
learn from experience, that whenever they occurred,
time ought not to be fpent in the frivolous preparations, which are ufually thought neceffary to precede
the giving of bark. If we judge it not to be proper
in any cafe to venture upon the ufe of bark, before
the body has been fufficiently prepared by emetics and
cathartics, we fhall too often meet with inftances
where the courfe of the difeafe will be finifhed before thefe preparations are completed. It has happened oftener than once to myfelf, that the fatal paroxyfm commenced before I had gone through the
ufual routine of preparation, which at that time I
thought indifpenfably neceffary, before I attempted
to cut fhort the courfe of the fever by means of its
well known fpecific.

Having premifed fo much with regard to the ge-

neral nature of the difeafe, I fhall now add a few obfervations about the management of thofe remedies
which have been moft ufually employed; fome of
which appear to be fuperfluous; while there is only
one of them which has a right to be confidered as effectual. It is a common and obvious remark, that
the intermitting fevers of the fame feafon fhew a general tendency to run over a fimilar courfe, though
the modes of treatment may be fometimes directly
oppofite. As I had often taken notice of this fact
during the time that I remained in America, I began
to fufpect that the changes, which I had been accuftomed to attribute to treatment, were in reality owing
to a difpofition in the nature of the difeafe, which
was little affected by the ordinary prefcriptions. But
that I might in fome degree afcertain the truth of this
fufpicion, I felected thirty cafes of fever, which had
commenced within twenty-four hours of each other.
This experiment was made at Ebenezer, in the month
of May; where the difeafe, though highly epidemic,
had not yet difcovered any figns of malignity; fo that
danger was not apprehended from a delay of a few
days. I claffed thofe thirty cafes in three divifions,
without paying regard to the nature of the fymptoms
of any individual cafe. To one I prefcribed a repetition of emetics, at fuch intervals as were judged
proper; for another, cathartics of various kinds, or
managed in various manners; while I left the third
wholly to its own courfe. I watched the progrefs of
the difeafe attentively for the fpace of eight days, and
cannot pretend to fay, that I obferved any material
difference in the changes or appearance of thofe cafes,
which were treated in fo different a manner. The
type, which was generally fingle tertian at the commencement of the diforder, changed for the moft part
to double tertian or quotidian, after the fecond or
third paroxyfm; though not in a different proportion,
as far as I could judge, in thofe which were left

entirely to Nature, or which were treated in the manner which has been mentioned above. It deferves
however to be remarked, that the paroxyfms were
ufually milder, more regular and diftinct after the
repeated ufe of carthartics; as alfo that bark fucceeded
more fpeedily and more certainly after a repetition of
antimonial emetics.

Where bleeding is judged to be proper or neceffary in the cure of the intermitting fever, the circumftances are generally fuch as require that it precede
every other in order of time. Bleeding was often
found to be ufeful in particular cafes. It moderated
the violence of fymptoms, and feemed not to be without effect in removing a certain ftate of the fyftem,
which refifted the fuccefsful operation of the bark ;
but I muft likewife add, that there was feldom occafion to employ it in the fouthern provinces of America. It was blamed by fome as increafing the difpofition to relapfe: of this, however, I can fay nothing
from my own experience; and as I am difpofed to believe, that the lofs of a moderate quantity of blood,
feldom does material harm in this difeafe; fo I have
reafon to think, that it is feldom neceffary in warm
climates, particularly in the hot months of fummer or
autumn.

Emetics have been employed in the cure of intermittents for a long time paft. There are many practitioners, who confider there ufe as indifpenfable ; and
fome have pretended, that the cure of the difeafe, in its
earlier ftages, may be trufted to this remedy alone.
Emetic tartar, fo managed as to operate at the hour
of the fever's return, was fometimes found to prevent
the accefs of a particular paroxyfm ; but though
obliged, from want of bark, to have recourfe to this
method of treatment in numberlefs inftances ; my
experience does not fupply me with a fingle one,
where I could pofitively fay, that it abfolutely cut
fhort the courfe of the difeafe. Relapfes, I muft

T

confefs, difappeared frequently in confequence of the practice; but they likewife difappeared frequently, where nothing at all was done : fo that I cannot help being of opinion, either that the real effects of emetics have been miftaken, or that the proper modes of managing them are not generally known.  I muft not however omit to mention, that emetics are occafionally of great fervice, and that antimonial emetics efpecially obviate the effects of inflammatory diathefis, and on particular occafions facilitate the fuccefsful operation of the bark.  They are likewife feldom followed by thofe dangerous effects which frequently arife from the employment of them in fevers of Jamaica ; though I certainly fhould advife that they be ufed with great caution in the fouthern provinces, in the fummer and autumn, where the remiffions of the fever are obfcure.

Cathartics have likewife been much employed by practitioners in the cure of intermitting fevers ; and in few cafes, perhaps, can be omitted with fafety. They certainly poffefs very remarkable effects in rendering the form of the difeafe regular and diftinct ; but they do not feem to be endued with a particular power of cutting fhort its courfe.—The circumftances of the cafe can only point out the propriety or advantage of the various remedies of this clafs.

The above are the ufual modes of evacuation, which moft practitioners confider to be indifputably neceffary previous to the exhibition of bark. I muft however remark, that cafes fometimes occurred in the fouthern provinces, particularly in the hot months of fummer, which only admitted of thofe evacuations in a fmall degree.  Inftead of the diftinct and regular paroxyfms, which were expected to follow the ufe of emetics or cathartics, the difeafe was fometimes obferved to affume a languid and continued form, in confequence of fuch treatment, while it likewife in fome inftances recovered the diftinction of paroxyfm and remiſſion, by means of applications, which excited the tone and vigour of the ſyſtem; particularly by means of exerciſe in cooler air, or accidentally by expoſure to rain.

Peruvian Bark is the chief remedy, upon which we now depend, for the cure of intermitting fevers. It is a remedy, which like others, has undergone ſome reverſes of reputation, ſince its firſt introduction into Europe; and, though its efficacy is now fully acknowledged, perhaps over-rated by Engliſh practitioners, it does not ſeem yet to have gained the ſame general credit with other European nations. The French uſe it with caution, and many of the Germans are ſtill its enemies. It has been accuſed even by ſome of the Engliſh writers, of failing in the cure of intermittents; and blamed by many of occaſioning complaints more dangerous in their nature than thoſe it was intended to remove. I was early aware of theſe objections, and watched narrowly that I might diſcover its real effects; and am warranted in ſaying, that it has every right to be conſidered as a ſpecific in ague and fever; while it is totally free from the imputation of occaſioning dyſentery, dropſy, or viſceral obſtruction. Thoſe complaints were always moſt frequent, where this remedy had been the moſt ſparingly employed.

But though I have mentioned that bark is both a ſafe and efficacious remedy in the cure of intermitting fevers, I muſt alſo confeſs, that it is only rendered ſo by particular modes of management. It is probable that much of the bark, which is now imported into England, is either in ſome ſhape adulterated, or naturally inferior in quality to what it had been in former times; as we often read of cures effected by a ſingle drachm in the laſt century, which we ſhould ſcarcely now expect from an ounce. Three or four ounces ſeldom failed of checking the progreſs of the moſt formidable fevers of America; one or

T 2

two frequently did not produce any fenfible effect. Being perfectly convinced of the truth of this obfervation, I generally gave bark in cafes, where the circumftances were judged proper for its exhibition, in dofes of two drachms ; which I directed to be repeated every two hours during the abfence of the fever. By this mode of treatment the difeafe was often fo completely conquered, that the patient was frequently capable of returning to his duty in the fpace of feven or eight days. Time ought not to be fpent in frivolous preparations, or difeafes attacked with feeble remedies, where the health of foldiers is concerned. The fpeedieft cures are generally the beft ; and I have had many opportunities of witnefling more health and ftrength gained during eight days in the field, than I fhould have expected from a month's cafe in an hofpital, affifted by the beft advice of the phyficians. The above-mentioned quantity of bark, for the moft part, was fufficient to effect a cure in ordinary cafes of the difeafe ; but where danger appeared to be threatening, the dofes were often increafed to half an ounce, or even more. In fome cafes of obftinacy, indeed, accompanied with a fluggifhnefs of conftitution, I obferved no other rule in the quantity than fuch as proved difagreeable to the ftomach, or excited a tumult in the fyftem. The method was often fuccefsful ; and I may obferve in general, that two ounces taken at five or fix times, and in the fpace of eight or ten hours, were often more effectual, than double the quantity in fmall dofes, and at long intervals. If the large dofes of bark, which I have recommended, fhould appear to any one to be dangerous or unneceffary ; I may add, that I have myfelf frequently taken an ounce at once ; while I have likewife obferved the cure to be tedious and uncertain with the moderate dofes of ordinary practice. I may farther remark, that this remedy was often rejected by the ftomach, and in fome cafes paſſed off almoſt inſtantly by ſtool ; yet that the courſe of the fever ſeemed to be no leſs effectually checked by it, than when ſuch effects did not occur.

The quantity of bark, which I frequently pre-ſcribed in the intermitting fever of America, may appear to be greater than neceſſary : ſo the time at which it was ſometimes given, may alſo appear to be premature. Bark was ſeldom given in the fevers of the ſpring and beginning of ſummer, unleſs in caſes of relapſe, till after ſufficient evacuations had been premiſed. In the autumnal months, where ſigns of malignity and danger were diſcovered, the firſt inter-miſſion was often laid hold of : neither was it uni-formly deemed neceſſary, to premiſe the evacuations of vomiting or purging. In relapſes I ſeldom let paſs the opportunity which the firſt intermiſſion af-forded ; by which means, thoſe who were ſubject to the returns of this diſeaſe, were rarely returned in the reports of the ſick.

Where bark was given in ſuch quantity, and in ſuch manner as I have mentioned above, it generally was ſucceſsful in checking the progreſs of the diſeaſe ; yet I muſt not omit to obſerve, that inſtances ſome-times occurred, where it totally failed of this effect, under every mode of management that could be de-viſed. It would be uſeful to aſcertain thoſe circum-ſtances exactly; but this is a taſk which I cannot promiſe to perform. It was however remarked, that where the inflammatory diatheſis prevailed in very evident degree, bark was frequently given without ſucceſs : and owing to this cauſe, perhaps, it was leſs to be truſted to in the ſpring, (unleſs in caſes of relapſe), than in the hot weather of ſummer and au-tumn. But beſides the obvious marks of inflamma-tory diatheſis, there appeared to be other conditions of the frame unfriendly to the ſucceſsful operations of bark. The exact nature of theſe, indeed, was perceived with difficulty ; yet I have often obſerved

them to be connected with fome of the following circumftances; viz. with a fmall and hard pulfe, or with
a pulfe where the ftroke was obfcure or without expanfion, and where a creeping or vermicular motion
was obfe, ved in the ftate of the artery; to which
was fometimes added, a conftricted ftate of the fkin,
a clammy moifture on the furface of the body, without figns of free perfpiration, and together with a fufpenfion or irregular action of the nervous influence.
In the ftate defcribed above, ¬bark alone was often
given in great quantity, without producing any fen-
ble effect. It fometimes fucceeded where antimonials, opiates and other antifpafmodics were joined
with it; but blifters applied to the back part of the
head and neck, were ferviceable above all other remedies in removing thofe circumftances, whatever
they were, which ftood in the way of its fuccefsful
operation. In fome cafes which had proved obftinate
to every other means, the difeafe difappeared immediately after their application; and in every one
where they were employed, ceafed any longer to refift the bark. Bark alone undoubtedly has a right
to be confidered as a fpecific in the cure of the intermitting fever, but its virtues are occafionally improved by the addition of aromatics, chalybeates, and
particularly by a certain proportion of fnake root.
It was a practice with fome of the country people of
Carolina, to attempt the cure of the intermitting
fever by means of Virginia fnake root, given in dofes
of two fcruples or half a drachm. I made a trial of
that remedy in feveral inftances; but did not find
that it was fuccefsful. Joined however with the
Peruvian bark, in the proportion of two drachms to
an ounce, it was often obferved to produce very excellent effects. The cures were more complete and
more permanent. There was not only lefs tendency
to relapfe, but dyfenteries and dropfies were more
rare, after I was fortunate enough to adopt this mode

of practice, than they had been during the preceding years.

There have been many different conjectures about the mode of the bark's operation in the cure of intermitting fevers; but none of them afford a satisfactory explanation of the subject. Bark is evidently bitter, aftringent and aromatic; but how it becomes fpecific in intermittents, in a degree fo fuperior to all other bitters, aftringents and aromatics, is a myftery we cannot eafily comprehend. It obvioufly poffeffes a power of giving tone and vigour to the powers of life; and often communicates to the fanguiferous fyftem, a certain ftate or difpofition, known by the name of inflammatory diathefis. Thus it is ufually obferved, that where the nervous frame is weak and delicate, bark rarely failed of cutting fhort the courfe of the difeafe, and its fuccefs in fuch cafes, is preceded, for the moft part, by apparent changes in the general diathefis of the fyftem. If ftrength, fulnefs, and vigour of pulfe follow the employment of bark, the fever frequently difappears; but if thofe figns are wanting, it is not poffible to form any certain judgment of the future effect. In like manner, if figns of inflammatory diathefis continue to prevail during the courfe of the difeafe, bark fometimes changes it to a continued fever; but feldom checks its progrefs effectually. From thefe facts, which I have often feen verified, I am difpofed to conclude, that bark is only to be confidered as an accidental fpecific in the cure of intermittents, and that its falutary effects probably may be explained from the change, which it occafions in the relative ftate of the body. That bark is fo often fuccefsful depends, we may prefume, on the intermitting fever being fo often connected with an attonic ftate of the fyftem; from which caufe it probably arifes, that it is fo much more effectual in fummer and autumn, in warm weather and in warm climates, than in the oppofite circumftances. But though an attonic ftate of the fyftem is frequently connected with intermitting fever, it is not conftantly fo ; hence the difeafe is not invariably removed by thofe proceffes which excite tone, or give rife to the inflammatory diathefis. Exceffive evacuations and other caufes, by which the body has been reduced to the laft ftate of debility, have often interrupted the courfe of an intermittent. So that we have every reafon to conclude, that bark is only relatively fpecific, in as much as it deftroys certain circumftances of aptitude, whch are effentially connected in particular fituations with the exiftence of the difeafe. In this manner, voyages, journeys, new purfuits, or new modes of life, frights, the active fervice of the field, or the hardfhips of fieges have often removed intermitting fever, which had refifted the ordinary aids of medicine.

I obferved in a former part of this treatife, that it was feldom a matter of much difficulty to ftop the courfe of the intermitting fever of America ; but I muft now add, that it was always difficult, fometimes impoffible to fecure the patient againft any future return of the difeafe. Bark, though much celebrated for this purpofe, did not feem to prevent a relapfe with the fame certainty with which it ftopped the courfe of the fever, when actually prefent. It deferves however to be remarked, that where bark had been given at regular intervals after the difeafe difappeared, the paroxyfms in the relapfe were, in general, not only flighter, but had ufually more of the inflammatory diathefis joined with them, while the complaint fhewed more difpofition to terminate of its own accord, after a few revolutions. It is univerfally known, that the powers of bark feldom fail in the cure of intermitting fevers, where given in fufficient quantity ; yet I muft alfo obferve, that its virtues do not feem to extend farther than to a temporary fufpenfion of the paroxyfms. That bark does not elimenate or deſtroy the actual cauſe of the diſeaſe, appears plainly from this fact, that relapſes are frequently the conſequence of thoſe circumſtances, which occaſion debility, or which counteract the effects of this tonic remedy. To which we may add, that though relapſes are often of a different type from the original fever, yet, as they generally happen on an even day from the ſuppreſſion of the paroxyſm, there can be little room to doubt that the old complaint again reſumes its courſe, though it probably, in the mean time, loſes ſeveral of its original ſymptoms. It is a fact likewiſe which we ought not to omit mentioning, but which in general, does not ſeem to be much attended to, that ſome periods are more remarkable for the relapſe of intermitting fevers than others. I obſerved before, that relapſes almoſt conſtantly happen on the even days. I now add, that the moſt remarkable of theſe days are the ſixth, the eighth, twelfth, fourteenth, twentieth, twenty-ſecond, twenty-eighth, and thirtieth. The fourteenth is remarkable for relapſes above all the others. Next to it we may rank the twelfth, twentieth, and twenty-ſecond; unleſs in times of very prevailing ſickneſs, where the ſixth and eighth often come in for a great ſhare. If we take pains to examine the particular circumſtances of the patient, and attend to the nature and degree of the prevailing epidemic, we may often be enabled to form a tolerable conjecture with regard to the moſt probable period of return. Having therefore acquired from obſervation ſome general ideas of the different propenſities to relapſe in different ſituations, and in different ſubjects, I uſually began to give the bark in quantity, and to uſe other precautions on the fifth, after the ſuppreſſion of the paroxyſm, in caſes where there were the ſtrongeſt ſuſpicions of a ſpeedy return; while this was delayed till the eleventh, nineteenth, or twenty-ſeventh in others, in proportion to the different degrees of healthineſs. This practice was

continued for the space of three days, or till the suf-
picious period was paft. But I muft further remark,
that befides the propenfity, which was obferved in
fevers to return at the periods above-mentioned, the
approach to the new and full moon was likewife found
to be connected with relapfes in a very remarkable
manner. Independent therefore, of the precautions,
which were ufed at the feptenary periods, the approach to new and full moon was conftantly attended to.

If the above directions were carefully complied
with, we might in general prevent the difeafe from
proving fatal, or from injuring the conftitution materially, though I muft at the fame time confefs, that
we could in reality do little more than preferve the
patient in a valetudinary ftate, till cool weather, a
change of fituation, or fuch a change in the manner
of life as excited the active powers of the conftitution, contributed their part to effect a permanent
eftablifhment of health. Being perfectly convinced
of the truth of this obfervation, I generally remitted
the men to the regiment, to be put upon the lift of
duty, as foon as I was certain that the courfe of the
difeafe was actually ftopt. The practice at firft fight
may appear harfh ; but I have found it to be falutary.
Exercife, even fome degree of exertion, promotes
the recovery of health. Habits of floth and indolence are fpeedily contracted in hofpitals : the military
ardour is gradually extinguifhed, and bodily ftrength
is recruited more flowly than in the field, under every
difadvantage of fatigue or inclement weather.—Of
the truth of this I have had ample experience.

I have now pretty fully defcribed the method of
cure, which I purfued in the intermitting fever of
America. I am not ignorant that other plans have
been adopted, and other remedies employed by others ;
but as I have not had experience of any, except that
which I have mentioned, I do not reckon myfelf

qualified to furnish any remarks on the subject. Bark, indeed, is so safe, and at the same time so effectual, that I should be slow in recommending any other remedy, where this can be procured in sufficient quantity. The strongest proofs of its value arise from a comparative view of the mortality of the intermitting fever, in different regiments, which were employed on the same service, but which were treated in different manners by their respective surgeons.— The Hessians were all of them inveterate enemies to the bark ; and there were ever some of the British surgeons who employed it very sparingly. The mortality among the troops trusted to the care of those, was uniformly in great proportion. There was a Hessian regiment, the situation of which I had the opportunity of knowing exactly, that lost one third of its men by this disease and its effects, during one year's service in Georgia. There were British regiments also, which lost more than a fourth ; while there were others, which did not lose a twentieth. The whole of these regiments were engaged on the same services ; they were all alike foreigners in America ; and there appeared to be no obvious cause for so great a difference in the degree of mortality, except a difference in the management of the bark. Bark was scarcely ever employed in one case ; in another it was used with timidity ; whilst it was given with the earliest opportunity, and in quantities far exceeding the usual practice in the third.

I have described, in the preceeding pages, the method which I adopted in the cure of the intermitting fever of America, whilst that fever preserved its distinct and proper form. I have likewise mentioned the best means I am acquainted with, of guarding against its return ; and it will not be improper in the next place, that I add a few remarks on some of its most usual and formidable effects. The intermitting fever of America shewed a strong disposition to

change into a fpecies of dyfentery, or a purging
and griping at particular feafons of the year, and more
efpecially in particular local fituations. Dropfy was
likewife a frequent effect of this complaint, and obftructions of the vifcera were not by any means uncommon, where the intermittent, from neglect or
other circumftances, had been allowed to go on in an
uninterrupted courfe.

The changes from intermitting fever to dyfentery,
and from dyfentery to intermitting fever, were fo frequent in the months of Auguft and September, that
thofe difeafes feemed evidently to depend on the fame
general courfe; affuming at different times the one or
the other form from caufes which we could feldom afcertain. In thofe cafes of dyfentery the ftools
were uncommonly copious and watery, and remiffions
and exacerbations frequently appeared at regular periods; but though the cure was often attempted by
the bark, it did not in general fucceed.

I muft begin with acknowledging, that I fhall not
be able to give a complete or accurate hiftory of the
progrefs and final termination of this fpecies of dyfentery, into which the intermitting fever is fo much
difpofed to degenerate; as I fhall likewife only have
it in my power to mention the general methods of
cure, which were purfued in the earlier ftages of the
difeafe. Where it was found that a cure could not
be accomplifhed in a reafonable time in the field or
regimental hofpital, dyfenteric patients were generally
removed to places where they could meet with better
accommodation; fo that the difeafe in its latter ftages
has feldomer fallen under my obfervation.

I muft obferve in the firft place, that this fpecies
of dyfentery had no right to be confidered as an in-
fectious difeafe. It appeared in fact to be no more
than an intermitting fever, which, from fome caufe or
other excited its principal force on the alimentary canal. Remiffions and exacerbations were generally

obfervable in the one difeafe, as well as in the other in the earlier ftages; yet thcfe appearances became gradually lefs and lefs remarkable, and there appeared at laft marks of permanent affeCtion of the inteftines. The fkin now became dry and harfh, the flefh wafted, and the bowels were uncommonly irritable, particularly where the *prickly heat* had retired from the furface; the difeafe was now evidently fupported by the exiftence of permanent local affeCtion.

With regard to the cure of this fpecies of complaint I have little to obferve, which is not generally known. Bark was fometimes employed to check its courfe; yet I muft confefs, that bark alone was feldom found to be fuccefsful. Where there were no marks of an actual inflammatory ftate of the ftomach and bowels, it fucceeded better when joined with aromatics, powder of camomile flowers, and particularly with fnake root. Laxatives were prefcribed frequently, and feemed often to be proper. They were however more ufeful where fome other thing was joined with them, which had the quality of determining to the fkin. Opium in various forms was a common remedy; and often a ufeful one before there were marks of permanent affeCtion of the inteftines. Ipecacuanha, or fuch preparations of antimony as promoted the evacuations by the fkin, were frequently combined with it. The great object which I purfued in this difeafe was to reftore and fupport a free perfpiration, to diminifh the irritability, and to ftrengthen the tone of the alimentary canal. I was difpofed to expect benefit from warm bathing, frictions, &c. but the fituation in which we were placed did not admit of a trial of them. Exercife was proper, and even fome degree of exertion. Change of air was ferviceable in many cafes, well feafoned food in fome, and wine in others. The above plan was purfued in the beginning of the complaint with tolerable fuccefs; but if it failed, or could not, from the

U

circumftances of the fervice, be properly executed, the difeafe then loft the remitting form, the body became lean and exhaufted, the ftools bloody, with a very irritable ftate of the bowels. In fuch cafes there was often ulceration, various degrees of inflammation, or obftrudtions in the coats of the inteftines. Blifters applied to abdomen or loins, and kept open, were often ferviceable; medicated glyfters, varied according to the nature and feat of the affedtion, were likewife of benefit; and in the latter ftages I have feen much good from the employment of ftrong aftringents. In illuftration of this I fhall mention the cafe of an officer, who was attacked with this fpecies of dyfentery, foon after the fiege of Savanna. Every thing, which the medical people of the garrifon could fuggeft was put into execution, without much benefit. The difeafe continued through the whole of the winter without material abatement ; the flefh wafted, the fkin became dry, with fuch other fymptoms as are ufual in this complaint. In the month of March, a perfon of the country recommended the decodtion of the bark of a tree, (probably of the fpecies of the Simarouba,) which appeared to be poffeffed of a confiderable fhare of aftringency. It checked the purging and griping inftantly ; fo that the difeafe ceafed for the fpace of three weeks. The appetite was good, the ftools copious, and in fome degree lienteric. The griping and even purging at laft returned ; the decodtion was repeated, but had not the fame effedts. Other aftringents, the extradt of logwood, terra japonica, &c. checked it for a fhort time, but no material ground was gained. He died in the month of May.

Dropfy fucceeding, and fometimes alternating with intermitting fever, was not by any means a rare appearance in America, particularly in fome feafons, and in fome fituations. The fwellings generally prevailed in every part of the body. They were ufually leucophlegmatic, though I have alfo feen fome inftances of tympanitis; a difeafe which was commonly fuppofed to proceed from improper management of the bark.

With regard to the cure of dropfical complaints, I have little to obferve which is not generally known. It confifted not only in evacuating the water, but in communicating to the fyftem fuch a degree of tone and vigour as refifted farther accumulation. With this view exercife, and even fome degree of exertion, was proper: wine, even more ftimulating liquors, high feafoned food, frictions and warm clothing, were ferviceable. I fhould likewife have been difpofed to expect benefit from fea bathing; but I cannot venture to fay, that I have ever made a proper trial of it. Blifters were alfo ufeful; not only as occafioning a difcharge of the waters, but as exciting the action of the vafcular fyftem. Among the numerous clafs of diuretics, there is not any one, which has fo powerful effects as cantharides, in fubftance or in tincture; and among the corroborants, I fhould be inclined to give the preference to chalybeates, colombo root, and Peruvian bark. It may not be improper to mention in this place, that I have feen fome inftances where a general anafarca has been completely cured by the accidental fupervening of convulfions.

Obftructions of the abdominal vifcera are likewife reckoned among the common effects of intermitting fevers. They are frequently attributed to the early or improper ufe of bark; but appear in fact generally to arife from the long continuance of the difeafe. I cannot add any thing to the manner of treating them that is not known to every one.

# CHAP. XIII.

HAVING defcribed the method of cure, which
I followed in the remitting fever of Jamaica,
the yellow fever and intermitting fever of America;
it will not, I hope, be deemed fuperfluous to give a
fhort fketch of the general principles, which have directed the practice of phyficians in febrile difeafes,
from the days of Hippocrates, till the prefent times;
a fubject of which I have not yet feen a connected
view. There is reafon to believe that the fcience of
medicine made confiderable progrefs in different
parts of the world, particularly in Egypt, at an early
period; but diftinct records of the art, prior to the
age of Hippocrates, are either loft, or fo blended
with the writings which are affigned to that author,
that we do not know how to diftinguifh them. The
practice of Hippocrates in fevers, has been accufed
of being feeble and inert; and it is certain, that
many of the moft effectual remedies of modern times,
were unknown to that venerable phyfician; yet if
we take the pains to eftimate candidly the whole mode
of proceeding, we fhall be forced to acknowledge,
that its effects were lefs inactive than has been generally imagined. The mode of treatment adopted by
the Coan Sage, feems not only to have done evident
good, but fometimes to have actually cut fhort the
courfe of the difeafe. Sweating at an early period,
the moft effectual means we yet know of cutting off
the courfe of fevers, was employed frequently by this
author; and though antimonials were then unknown,
yet fweat, excited by a fimpler procefs, was often
obferved to produce very decifive effects. Hippo-

REVIEW OF PRACTICE, &c.    221

crates, indeed, has been lefs circumftantial in the
detail of remedies than we could have wifhed.    He
has however related the dietetic part very diftinctly.
His rules are always judicious, and his regimen
fometimes of fuch efficacy, as evidently to operate
very confiderable changes in the ftate of the fyftem.
It appears frequently to have been his aim to attempt to exterminate the fever in its early periods,
or to endeavour to cut fhort its courfe abruptly, by
direct or indirect oppofition to its proximate caufe :
yet this idea was not purfued beyond a certain point.
After the fourth day had paff.d, Hippocrates ufually
contented himfelf with fupporting the general powers
of life with proper diet and nourifhment, allowing
nature after that period to perform the work her own
way: in other words, to complete the bufinefs by the
flower operations of coction and crifis.    Thefe two
views, viz. the attempts to cut fhort the difeafe
abruptly in its beginning; or in the late periods, the
endeavours to fupport the powers of life, till the natural termination might arrive, comprehend the general rules of practice followed by this celebrated phyfician : and I much doubt if the moderns, notwithftanding all their pretenfions, have actually difcovered any other indications more decifive, though I
willingly allow, that they have made great improvements in the mode of executing thefe I have mentioned.    At leaft, I acknowledge for my own part,
that I am not yet acquainted with any mode of treatment, by which the natural courfe of a continued, or
even obfcurely remitting fever can certainly be prevented, after the firft days of the difeafe are paft ;—
that is, after a diftinct formation of the type; before
that happened, the fweating procefs is frequently fuccefsful.    But though the attempt to cut fhort the
dif fe in its beginning, or to fupport the powers of
life in the later periods, comprehend this author's
general rule of practice ; yet we find confiderable

U 3

diverfity in the manner of accomplifhing thefe different purpofes.  The παντα παντως, or an attempt
to counteract the derangements of morbid caufes,
may be confidered as the firft general maxim, which
was eftablifhed in the cure of difeafes.  Hippocrates
has this maxim conftantly in his eye, and endeavours
by various means, according to a fuppofed diverfity
in the mode of action, to cut off the immediate exiftence of fever.  In this manner the prefence of heat
and bile, or the fuppofition of obftructed perfpiration,
have feverally furnifhed him with different indications.  His ideas however are not precife; fo that
his practice frequently fluctuates between conjecture
and experiment.  If the difeafe does not yield to one
mode of treatment in a given time, he frequently
paffes to its oppofite much at random.

The doctrines of Hippocrates acquired fuch general credit, in every part of the world where they were
known, that we do not remark any material innovations in the cure of fevers, till the time of Erafiftratus; a fpace of near two hundred years.  The intervening period, indeed, was diftinguifhed by philofophers, who employed their time in inveftigating the
ftructure and economy of the human frame, as well
as by phyficians, who improved the art of medicine
by the invention of new remedies.  Plato, Ariftotle
and Theophraftus were the moft eminent of the former; Petro is chiefly diftinguifhed among the latter.  (1) This author, as we find recorded both by
Celfus and Galen, attempted to extinguifh a fever
by copious drenching with cold water; foon after
which, he nourifhed his patient with wine and ftrong
foods, a cuftom which was in fome degree imitated
by Clophantus.  Hints, however, of the firft of thofe
practices are found in the writings of Hippocrates,
fo that we may juftly confider Erafiftratus, as the firft
who departed fo far from the principles of the Coan
Sage, or who arrived at fo high reputation,  as to be

regarded by pofterity as the author of a new method of curing difeafes. I mentioned in a former part of this treatife, the opinion which Erafiftratus entertained concerning the caufe of fevers. All that we know of his practice may be comprifed in a few words. Erafiftratus was every where the inveterate enemy of bleeding. He was likewife the author of a certain plan of abftinence, which, with a little modification from Afclepiades and Themifon, made a confpicuous figure in the annals of phyfic for feveral fucceeding ages.

Herophilus, who lived much about the fame time with Erafiftratus, acquired alfo high reputation in the medical world ; but unfortunately we have it not in our power to give a particular detail of his difcoveries. Soon after the period I have mentioned, the fcience of medicine was divided into three diftinct branches ; viz. as the art is employed to remove difeafes by diet, by drugs, or by manual operation. Serapion likewife, who is confidered as the author of the empiric fect, made his appearance not long after this divifion of the profeffion into feparate branches. His followers were numerous, and many of them were refpectable; but we are not enabled to give a diftinct account of their practice on the fubject of fevers. The fummary views of Celfus and Pliny, or the accidental fragments in the voluminous works of Galen, furnifh only imperfect information. Contentions, however, ran high between dogmatics and empirics, the former of whom were chiefly guided by reafonings, as the latter trufted folely to experience.

The Greeks, for many ages, were the only people we are acquainted with, who cultivated the fciences with induftry. For near five hundred years they were almoft exclufively the profeffors of the healing art. The Romans were cautious of admitting the refinements of learning into their ftate; and had nearly attained the height of their glory, before they willingly received phyficians into their city. Afclepiades the Bithynian, the contemporary and friend of Pompey and Cicero, appears to be the firft, who practifed medicine at Rome with any degree of reputation. This author left his native country, with the defign of inftructing the Roman youth in the arts of eloquence ; an acquirement, which was held in high eftimation among that warlike people; but not fucceeding in this purfuit, according to his wifhes, he foon difcovered, that the profeffion of medicine offered a fair opening for the exercife of his talents. The ftate of the art, indeed, was then peculiarly propitious to his undertaking ; the former fame of Hippocrates being divided between Erafiftratus and Herophilus, and fome authors of the empiric fect. Afclepiades was two difcerning not to perceive, that new doctrines could not fail of drawing followers, and too enterprizing not to attempt to carry this purpofe into execution. He probably, in the firft place, read over all that had been written by former phyficians ; the moft effential parts of which, he felected with a good deal of art, and fo modified as to form a fyftem, which appeared to be complete, and which exhibited upon the whole, confiderable appearances of novelty. If we take the pains to trace his opinions to their fources, we fhall find that Democrites or Epicurus furnifhed him with his philofophy, that Herodicus fupplied him with the hints of bathing, friction, and geftation, that the plan of abftinence, or fafting for three days, was learnt from Erafiftratus ; and that Petro and Cleophantus inftructed him in the practice of giving cold water, and of allowing greater indulgence in the ufe of wine.— But though the authors I have mentioned furnifhed Afclepiades with the hints of his doctrines and practice; yet we may obferve, that he has always modelled thefe after his own fafhion, and carried them

farther than had ever been done by their original in-
ventors. It was remarked above, that the profeffion
of medicine was divided into different fects before
the time of Afclepiades ; yet, as far as we can judge
from the imperfect information which has been tranfmitted to us, this author was the firft, who deferves
the name of an active phyfician. Previous to the
time of this enterprizing Bythynian, medical men
trufted the cure of fevers chiefly to the efforts of
nature; and were only folicitous about the mode of
death. Afclepiades, affuming a bolder principle, endeavoured to cut fhort the fever in the midft of its
courfe : and it is reafonable to fuppofe, that by ftepping fuddenly from the extremes of torture to the
higheft gratification of the appetites, fuch changes
were effected in the ftate of the fyftem, as probably
fometimes accomplifhed the purpofe. Such were
the general views of this author. It may not be improper to add a few remarks on fome of the remedies
which he employed.

It is impoffible to fix the date, when friction and
warm bathing were firft numbered among the affift-
ances of the phyficians. They appear, however,
from the teftimony of Celfus, to have been ufed with
caution by the ancients. Afclepiades not only indulged in them with freedom, but fo conducted the
management of bathing, that it might be juftly confidered as a luxury of the higheft elegance. Geftation, another of the gymnaftic remedies employed by
this author, even in an early period of ardent fever,
is an experiment of a ftill bolder and more defperate
kind; and fuch as fuccceding writers with one voice
feem to have condemned :—perhaps without examination or fair trial. I am fenfible that the opinion,
which I am to offer on this fubject, is not likely to
meet with general approbation ; but the opportunity
which I enjoyed, during the late war, of ferving
with a regiment, which was almoft conftantly in the

field, enables me to confirm the truth of it by ample experience. The good effects of geftation or travelling, even in awkward conveyances, were very obvious in almoft every ftage and fituation of the ardent bilious fever; but I fhall relate fome inftances of its fuccefs, which appear to preclude all poffibility of doubt. At Ebenezer in Georgia, at a feafon, when the thermometer, in the coolelt part of the houfe, often ftood at ninety-fix, and even fometimes rofe above it, I was feized with the ardent bilious fever, which at that time made dreadful ravage among the troops. For fix or feven days I did not once fhut my eyes; my thirft was great, yet every fort of liquid, which I could procure, was naufeous; the diftinction of paroxyfm and remiffion was no longer perceivable; the pulfe, at the wrift, was neither uncommonly frequent nor ftrong; but the pulfation of the defcending aorta was fo great, as to fhake the whole frame; anxiety and reftlefsnefs were intolerable: in fhort, the torment was fo exceffive, that human nature could fcarcely fuffer more. The fituation was precarious; and without much reflexion I indulged the defire of being carried to Savanna; though the diftance was not lefs than twenty-five miles. An open carriage, the only conveyance which the country afforded, was provided for the purpofe; and I was put into it, in a very feeble and diftreffed condition. Fortunately the day was cloudy, and cooler than ordinary. The roads were likewife foft and fandy. Though the carriage was very defective, the motion was no ways unpleafant; and I had not travelled two miles before I felt a wonderful increafe of vigour. It rained heavily about half way, and before I reached Savanna, I was drenched to the fkin. The effects which might have been expected, did not follow. Inftead of being hurt, I was furprizingly benefited. I walked into the houfe with ftrength and firmnefs, eat fomething without diflike,

and flept found the following night; in fhort, obtained
a perfect remiffion of the fever. This is a ftrong
inftance of the good effects of travelling in fevers ;
yet it is only a folitary one. I fhall add another,
which places the fact on a ftill firmer bafis. I mentioned in a former part of this treatife, that while the
71ft regiment lay at the Cheraws, the endemic of
the country prevailed among the men in a moft unufual degree. The difeafe was often without diftinction of paroxyfm and remiffion; the anxiety and
reftlefsnefs were intolerable, bilious vomitings and
purgings were frequent, and exceffive. While near
two hundred men were in this fituation, an order
arrived for abandoning the poft. It being impoffible,
as we were fituated, to provide waggons to tranfport
fo great a number of fick, about forty of thofe who
were leaft likely to be foon fit for fervice were fent
down the river in boats. Of the particular fate of
this party I cannot fpeak with certainty ; but I have
the fatisfaction to add, that not a man died of thofe
who retired to Camden by land ; and that after the
third day, fcarcely a fever was left, which had not
affumed a regular intermitting form. This appears
at firft fight almoft an incontrovertible proof of the
falutary effects of geftation; but I, muft not at the
fame time omit to mention, that benefit alfo probably
refulted from a change, which accidentally happened
in the ftate of the weather. The weather, (which,
during the time we remained at the Cheraws was
uncommonly hot,) became unexpectedly cool after
the march was begun ; together with rain, from
which the fick men had nothing to fhelter themfelves.
This inftance of expofure to rain furnifhes a proof of
a fact of much importance. It is generally believed
that getting wet with rain is hurtful to a perfon in
health. It is no lefs commonly fuppofed to be certainly pernicious in ficknefs ; but the contrary appears to be fometimes the cafe. I have feen the
happieft effects from the application of cold, even

from getting accidentally wet with rain in many instances, besides the present. Increase of tone and
vigour was generally the consequence ; and life was
evidently protracted, sometimes perhaps saved, by
accidents, or modes of treatment, which, in the common opinion of mankind, would have been reckoned
the causes of death.—But though I have mentioned
the accidental good effects of gestation, and even of
exposure to rain in different states of the ardent fever ; I shall not be so paradoxical, as to recommend
such experiments in common practice. I must however be allowed to observe, that we have little cause
to be afraid of their pernicious effects. Motion and
travelling, as far as my experience goes, were constantly hurtful in cases of local pain and inflammation ;
or in derangement of intellect ; but, on the contrary,
constantly serviceable in anxiety and restlessness, depending on the state of the stomach ; as also in affections of the biliary system.

I mentioned before, that Asclepiades practised medicine at Rome with great reputation. He propagated his doctrines with a good deal of success ; but
such is, and ever has been the fate of our conjectural
art, that no system has yet preserved its credit undi-
minished for any length of time. The views of physicians, with regard to diseases, had hitherto been various and complex; even fevers had been often considered, by the same person, as depending on different causes. Themison, a pupil of Asclepiades, attempted to remedy the perplexity which necessarily
arises from this instability of arbitrary conjecture ;
and endeavoured to reduce all the disorders, to which
the human body is liable, to two general classes ;—
viz. to those which arise from an increased degree of
stricture, or its opposite affection, preternatural laxity;
to which he afterwards subjoined some complaints,
which appeared to partake of the nature of both.—
The idea of simplifying diseases did not probably arise

in the mind of Themifon, till the latter period of his life; on which account, perhaps, the doctrines were left in fome meafure unfinifhed; and it is to Theffalus, who lived in the time of Nero, that we are indebted for completing the methodic fyftem, and for enlarging the bounds of its fame. Soranus likewife added to its credit. It is a misfortune, which we muft regret, that except Cœlius Aurelianus, there is not one of the many authors who were attached to the tenets of this fect, whofe works have efcaped the wreck of accident or time. This only remaining author appears to have copied, by his own confeffion, almoft literally from Soranus.

It does not belong to this place to enter deeply into the tenets of the fect, with regard to difeafes in general; but leads to views of fome importance in the theory and treatment of many diforders. It proceeds on the fuppofition of circulation in all parts of the body; and with a little latitude may be fuppofed to comprehend the affections of the animated folid or contractile fibre. Fever is confidered by the writers of this fyftem, as a difeafe of the clafs of ftricture; and if we had authority to add fpafmodic, we might believe the methodics had difcovered a very important phenomenon in the hiftory of febrile difeafes.— But this fome may think is granting them too much. If we take the trouble to trace their doctrine, refpecting the caufe of fever, to its fource, we fhall find that the hints of it are furnifhed exprefsly by Afclepiades, who perhaps borrowed his ideas on the fubject from Erafiftratus, or even from Hippocrates.— But as ftricture, affigned by the writers of the methodic fect as the caufe of fever, is not perhaps radically different from the obftruction of preceding authors; fo we do not find much material difference in their manner of conducting the cure. The followers of Hippocrates, Erafiftratus and Themifon proceeded equally on the idea of reftoring permeability in the

X

minuter canals of the fyftem; a purpofe which they conceived would be beft affected by certain proceffes of attenuation and relaxation. Thus Hippocrates diluted plentifully, and gave nourifhment only fparingly, during the firft days of a fever. Erafiftratus enjoined a general abftinence, Afclepiades prefcribed a term for the duration of the abftinence; while Themifon limited it fo rigidly to the fpace of three days, that the practice was diftinguifhed by the name of diatriton, as its followers were known by that of diatritarii. This idea of diatriton conftituted an object of much importance in the fyftem of the methodic phyficians. It directed all their movements, and is the only view, which can properly be called their own. The mode of application of the remedies of preceding authors was occafionally modified by this fect; but except that which I have juft mentioned, we do not difcover much that is fundamentally new.

It may not be improper in this place to take notice of the practice of cold bathing in fevers, which was introduced at Rome in the infancy of the methodic fect; and which afterwards acquired great celebrity in different parts of the world. The Emperor Auguftus, who for the greateft part of his life was afflicted with ill health, was at laft attacked with a complaint of fo obftinate a kind, that the ufual applications did not afford him any relief. Warm bathing and all that train of remedies had been tried in vain. The Emperor was fenfible of his defperate fituation; and his phyfician Antonius Mufa, baffled in all his attempts, ventured, though apparently at great hazard, to prefcribe the cold bath. The health of Auguftus was unexpectedly reftored by it; and the phyfician was highly honoured, and amply rewarded. It is a misfortune that we do not know the fpecific nature of the difeafe, under which the Emperor laboured; but we have reafon to conclude, from the circumftances which attended the cure, that it was a fever of a bilious remitting kind; perhaps complicated with catarrhal affection, and wasting of the body. But though the fuccefs of cold bathing, in the inftance I have mentioned, was much greater than expectation; yet the remedy did not long retain its credit. It was foon afterwards employed in the cafe of Marcellus, a youth of great hopes, and prefumptive heir to the empire; but the event proving unfortunate, it fell fuddenly into difrepute,—probably without fufficient caufe. We find however, that Celfus, in lefs than half a century, afterwards ventured to recommend it in a certain fpecies of the flow or hectic fever; though Charmes, a phyfician of Marfeilles, appears actually to be the firft who rendered the ufe of the remedy general. Galen employed it frequently with great freedom and boldnefs. The Arabians, particularly in peftilential difeafes, went ftill farther than the Greeks or Romans; and we prefume, from the fact recorded by Bufbequius, that it was fometimes prefcribed at Conftantinople, even fo late as the fixteenth century. I do not know that it has been often tried in Europe, fince the revival of the medical fciences in the Weft. I mentioned in a former part of this treatife, that I had employed it frequently in the fevers of Jamaica; I now add, that I have ventured upon it in the fevers of this country with fo great fuccefs, that I fhould expect the moft beneficial effects might refult from a proper management of it. (2)

We are indebted to Celfus, who lived in the time of Tiberius, for preferving many of the opinions and practices of preceding phyficians, which otherwife would probably have been loft. This author, not lefs remarkable for candour, than for the elegance and perfpicuity of his manner of writing, does not feem to have been blindly devoted to the tenets of any particular fect. He has favoured us with valuable extracts from the works of the moft celebrated dog-

X 2

matics; he has likewise taken notice of the moft remarkable opinions and practices of the Empirics, without omitting to mention the innovations of Afclepiades and Themifon. With regard to his particular merit as a phyfician, we may obferve that he every where difcovers an excellent judgment, and that his practice is generally decided without being rafh.

So great have been the ravages of time or accident among the writings of the early phyficians, that it is fcarcely poffible altogether to avoid error. in attempting to trace the various revolutions in medical practice, prior to the time of Galen. From that downwards we are enabled to give a more certain and better connected view of the fubject; there being few of the principal writers of this latter period, who have not been preferved entire. When Galen came firft to Rome, which was in the time of the Emperor Antoninus, the practice of medicine was chiefly in the hands of the followers of the methodic fyftem. The practice of diatriton, or abftinence for three days was then in high fafhion, not only with the profeffed pupils of Theffalus, but even with the few remaining adherents of Erafiftratus. Galen every where declares himfelf its inveterate enemy; and often tranfgreffes the bounds of liberality and decency, in his attempts to turn it into ridicule. His own endeavours are exprefsly exerted to revive and eftablifh the principles of Hippocrates; and to complete thofe parts, which the want of time or the want of experience of his mafter had left imperfect. He poffeffed uncommon fertility of genius, a great flow of language, and a judgment by no means deficient; yet, from a fophiftical fpirit of philofophizing, he frequently fo entangled his opinions with theoretical diftinctions, that his views are often uncertain, and fometimes embarraffing. The principle with which he fets out is directly to oppofe the actual exiftence of fever; he next recommends to remove, at leaft to avoid an increafe of thofe caufes which give rife to the difeafe. Thefe ideas are drawn from the writings of Hippocrates, and are fuch as no perfon will difpute : but, as the caufes of fever are fuppofed, both by Hippocrates and Galen, to be many and various, fo the indications of cure often require to be executed in different, and fometimes in directly oppofite manners. This neceffarily gives rife to confufion ; and entangles the practitioner in the mazes of doubt and conjecture ; to obviate which, as much as poffible, the learned commentator of Hippocrates has thought fit to divide fevers into three general kinds, viz. ephemeral, continued, and hectic or habitual ; the caufes of which he fuppofes to be fo little analogous to one another, as to demand particular management in the method of cure.

We look in vain for new views, or material improvements in the management of fevers, in the writings of thofe Greek phyficians who followed Galen. Oribafius profeffedly is no more than a collector of the opinions and practices of other men ; and Aetius, on the prefent fubject, does not afpire to much higher fame. There are, indeed, few of his obfervations, which may not be found in the volumes of Galen, or fome preceding writer ; yet he feems generally to have comprehended what he wrote. He digefted the knowledge which he found in books with care and judgment; and gives an arrangement fo clear and perfpicuous, that the perfon may derive information from Aetius, who would be overwhelmed and loft in the prolixity of Galen's difcuffions.

From thofe writers, however, who trod implicitly in the footfteps of Galen, we muft be allowed to feparate Alexander of Tralle, a phyfician who lived in the fixth century. This author wrote his book on fevers at a very advanced age ; and though the treatife perhaps does not contain many ideas, which may

X 3

not, in fome fhape or other, be found in the writings
of his predeceffors, yet the obfervations have the appearance every where of having originally arifen from
actual experience. The language, which is concife,
clear and perfpicuous, is wholly his own. The ambiguous circumftances of difeafes are more accurately
difcriminated than in any preceding work which has
defcended to the prefent times ; and though the manner of accounting for things may be fometimes erro-
neous, yet it has had little influence on the practical
indications, which are almoft unexceptionably judicious. As Alexander of Tralle wrote at a time of
life when fame muft have been indifferent to him,
and to a friend, whom he was more folicitous to in-
ftruct than to amufe with the fplendour and variety
of his learning, we have an additional caufe to give
our confidence to his obfervations. His manner is
candid and ingenious ; and the treatife before us may
be confidered by the practical phyfician, as the moft
valuable of the remains of the ancients. Judicious
cautions are every where interfperfed, and confiderable changes in the management of remedies are
fometimes attempted ; but the practice of giving
opiates in a certain ftate of fever is the only practice
of this author, which has any title to be called innovation.

Paulus is the next phyfician of note, who lived
after the days of Galen. He was born in the ifland
of Aegina, and travelled over many countries. It is
probable that he was fufficiently acquainted with
every difcovery, which had been made by his predeceffors ; yet Galen, on the fubject of fever, is the author whofe works he has principally followed. His
book on fevers, indeed, contains all the material doctrines and obfervations of that voluminous write r;
and thofe who dread the labour of encountering the
prolix and fophiftical difquifitions of the commentator
of Hippocrates, may find a very diftinct analyfis

of his opinions and practices in the treatife of Paulus Aegineta.

Having endeavoured in the preceding pages to give a fhort view of the methods which were ufually purfued by the moft eminent of the Greek phyficians, in the cure of fevers, it will be neceflary in the next place to take fome notice of the improvements of their immediate fucceflors, the Arabians. This tafk will be foon performed; the Arabians have not in reality opened any views in the curative indications of febrile difeafes, which were unknown to their predeceflors; or which require that we fhould fpend long time in endeavouring to explain them. The medical fcience evidently drew its origin from the Eaft; yet it was alfo foon reconveyed to the countries from whence it fprung, with improvements and additions from the genius of the Greeks. We learn from Herodotus, that Democedes, a native of Crotone, who had ftudied medicine in the ifland of Aegina, far excelled all the phyficians of the Perfian court, even fo early as the time of the firft Darius; though the court of this Prince probably could boaft of all the fkill, both of Affyria and of Egypt. Clefias fometime after was held in great eftimation by Artaxerxes; and the invitation, which was held out to Hippocrates by the Perfian monarch, indicates very clearly, that the Greeks, even then, were more famed for medical fkill than the inhabitants of the Eaftern countries. The iflands and fhores of the Mediterranean feem through the whole hiftory of medicine, to have produced the greateft number of phyficians. Crotone and Cyrene were famous for feveral ages: and Alexandria, at a later period, rofe into great celebrity. Students flocked to it from every part of the world; it was even neceflary that every one, who afpired to wealth or reputation in phyfic, fhould fpend fome time in this celebrated feminary. It was owing perhaps, in fome degree, to the vicinity of

this illuftrious fchool, that the province of Syria enjoyed at one 'time, a confiderable fhare of learning
and learned men. The works of the moft eminent
of the Greek phyficians were tranflated into the
dialect of the Syrian country, in the feventh and
eighth centuries; by which means they were probably, in fome meafure, propagated in the Eaft :
though we alfo are informed by Abulpharage, an
Arabic writer, who had preferved many curious
anecdotes of private hiftory, that the doctrines of
Hippocrates were planted in the Chorafan, at a ftill
earlier age, by the phyficians, who followed in the
train of Aurelian's daughter, who was married to
Sapores king of Perfia : nor is it improbable, that
thefe doctrines were ftill more generally diffufed
through the Perfian dominions, by the alliances of
friendfhip, as well as by the long wars, which were
afterwards carried on between the Greek empire and
the celebrated Khorrou Pawiz. But though the inhabitants of Syria and Irak were an enlightened nation, at an early period; their neighbours, the Arabians, who afterwards attained fo great a name in
fcience no lefs than in war, remained long in a ftate
of illiterate ignorance. Before the eftabiifhment of
iflamifm, there fcarcely was a native Arab, who could
either write or read. The little genius they poffeffed
was chiefly exerted in compofing veifes, or in colouring a rhetorical harangue. They appear, indeed, to
have acquired fome practical knowledge of the motions of the heavenly bodies ; and it is likewife
reafonable to fuppofe, that they had the fame fkill in
medicine, as is common to favage nations ; but there
is no reafon to believe, that they, as yet, had made
progrefs in the medical art, conlidered in a fcientific
view. Hareth, a native of Tayef, who lived in the
time of the prophet, and who feems to have been in
habits of intimacy with that fingular man, is the firft
of the Arabs, whofe name is recorded among the

phyſicians of the Eaſt. This perſon, who acquired
ſome knowledge of medicine at Niſabour, and other
places in the Choraſan, returned home after ſome
time, with great wealth, and no ſmall ſhare of fame.
He practiſed among his countrymen with much reputation; but how far he ſpread the light of ſcience
among them is uncertain. The Saracens advanced
rapidly in conqueſts and the eſtabliſhment of their
faith; but we do not hear any thing of their progreſs in the healing art, till the ninth century. Syrians
and Perſians, generally of the Jewiſh or Chriſtian
religion, laboured ſometimes for the warlike Arabs
in the ſervile occupation of curing diſeaſes, at leaſt
we do not know that any of the Saracens attained
much eminence in medical ſcience, till the tranſlations of Honain and his pupils laid open to them the
treaſures of the Greeks. We are ill qualified at this
period to judge of the merit of theſe tranſlations. But
if we may be allowed to form concluſions, from the
uſe which has been made of them, we ſhall not, perhaps, be diſpoſed to entertain a very high opinion of
their accuracy. In many inſtances, the later Arabian phyſicians have expreſſed the ideas of Hippocrates and Galen only very looſely; and in ſome few
caſes, perhaps, have not very clearly comprehended
their meaning. But, as the later Saracens were
ſeldom ſkilled in any language except their own;
the original tranſlators are probably alone blameable
for the whole of theſe miſtakes.

The medical authors, who wrote in the Arabic
language between the ninth and fifteenth centuries,
and, who ſtill lie concealed in the leſs acceſſible dreſs
of their native country, are almoſt innumerable: neither are thoſe, who have been introduced into the
common acquaintance of Europeans few in number,
or ſmall in volume. If I poſſeſſed a complete ſeries,
even of thoſe who are commonly known in Europe,
the examination I have entered upon might be drawn

out to a confiderable length; but as I have no hopes
of obtaining that foon, I fhall content myfelf with giving fome id.a of the Arabian fyftem of practice in
fevers, from the works of Avicenna, the moft eminent and beft known of the Oriental phyficians. An examination, indeed, of one of the writers of this nation
may, in a great meafure, render an inquiry into the
others unneceffary. Thofe, at leaft whom I have
feen, do not differ materially from one another; or
perhaps effentially from the Greeks who went before them. The canon medicine, the principal work
of Avicenna, exhibits a fyftematic view of the whole
art of medicine, theoretical as well as practical. I
have read over with care all that relates to fevers;
and though there is little, perhaps which may not ultimately be traced to Galen or Hippocrates; yet the
author has not copied fervilely from either of them.
He is more full and particular than the one; lefs prolix and tedious than the o.her. I muft however remark, that the diftinctions and divifions, which he
has attempted to introduce into the hiftory of fevers,
are not only unneceffary, but actually ferve to em-
barrafs the indications of cure. His general theories are thofe of Galen. In the general conduct
of the cure, he treads in the footfteps of the fame
mafter. He appears, indeed, to be more fearful of
the lancet; while he is not perhaps always judicious,
or confiftent with himfelf, in the manner of employ-
ing it. On the contrary he has admitted cool air rather more freely, and has perhaps carried cold drink
even to a bolder length, than had been done by the
Greeks. Cool air, cold drink, and even the external
application of cold, may be reckoned among the moft
effectual remedies in the fevers of hot climates; and .
this author has conducted the management of them,
in a luxurious, elegant and efficacious manner. But
though the works of Avicenna furnifh a general view
of the practice of the Arabian fchool of phyfic, it is
ftill in fome degree a defective one. As he has not

furnifhed us with a detail of the cafe of an individual,
we are not able to judge precifely of his powers of
difcerning the difeafe, or of his decifion in the manner
of treating it.

The medical fcience, which after the taking of
Alexandria was little cultivated by the flothful Greeks,
or barbarous nations of the Weft, fprungup ith new
vigour in the province of Syria, in Irak and Arabia;
and followed every where in the train of the Saracen
conquerors. Extending with their arms over the
northern coafts of Africa, it foon found its way into
Spain; and, even fo early as the eleventh cen ury,
was conveyed to Salernum in Italy, by Conftantinus
Africanus, a native of Carthage, who had lived long
in Afia, and who was well acquainted with the language and medical knowledge of the Orientals. The
Arabians were the firft who opened the fources of che-
miftry; they alfo made great improvemen's in the art
of furgery, and even defcribed fome complaints which
in earlier ages were not taken notice of but they departed but little ftom the fyftem of the Greeks in the
management of febrile difea'es. After the fall of the
Roman empire the genius of learning made no exertion in Europe for a very long period of time.
The native European flothfully acquiefced in the imperfect knowledge of Arabian writers, which was
obtained from the inelegant, and perhaps often un-
faithful tranflations of the Jews, who, for a confiderable time, were no contemptible profeffors of the
medical art. But, though fome part of the knowledge
of the Arabian phyficians was communicated, in this
manner to the nations of the Weft, in the eleventh
and twelfth centuries; yet a part of the fixteenth paffed over, before it was poffible to trace any marks of
improvement. Commentaries were written without
number; but, for many years, there fcarcely was an
individual in all the feminaries of Europe who dared
to think for himfelf. It has been cuftomary to date

the revival of fciences in the Weft from the taking of Conftantinople, by which the ftores of Greek lite-rature were in fome degree opened to the world. The language of Galen began then to be more generally underftood, and the writings of Avicenna fell rapidly into neglect; yet the advantage which accrued to medicine from the change, does not appear to have been great. The mind was exercifed in a wider field of learning; but it was ftill in chains to the authority of the ancients. The opinions of Galen and Hippo-crates were copied, recopied and commented upon by hundreds; but there were very few who ventured to ufe any judgment of their own. Among the moft cele-brated of the followers of Galen we may reckon Fer-nelius, Foreftus, Lommius and Sennertus, men of confiderable talents, but who were too fcrupuloufly devoted to the principles of their mafter, to open a new road in the practice of the art. This was re-ferved for Paracelfus, who early in the fixteenth cen-tury ventured to attack the opinions of his predecef-fors, and the authority of Galen. Paracelfus poffeffed a confummate fhare of affurance, together with know-ledge of remedies which were not generally known at that time. He acquired fome acquaintance with the chemical difcoveries of the Arabians, in the courfe of his various travels, and applied in practice what he had learned, on his return to his native country. He defpifed the authority of the regular phyficians, em-ployed remedies with great boldnefs, and often with fingular fuccefs. This fuccefs was even very pro-bably exaggerated by report; and there appear to have been many, who followed him implicitly; while others exerted themfelves in modifying and improving his ideas. Under this laft view we may rank Van Helmont, a perfon, who effected a very material innovation in the manner of curing febrile difeafes Van Helmont poffeffed confiderable learning; but difcovered, at the fame time fuch marks of warmth and enthufiafm of

genius, as diminished his credit with contemporary
and succeeding practitioners. The terms which he
employs, are sometimes ridiculous; and his reasonings
are frequently disfigured with fancy and whim; yet
his ideas are generally important, and often well
founded. The archæus of this author does not differ
materially from the sentient principle (τα ωρμωντα)
of Hippocrates; and perhaps comprehends the whole
idea of the vis medicatrix naturæ of the moderns.
Van Helmont proceeds to the cure of fever on the
important principle of exciting, or calling forth the
powers of life, to exterminate an offending cause; so
that we may actually confider him as the first, after
Afclepiades, who attempted to take the bufinefs wholly
out of the hands of nature. He difregards the pro-
cefles of coction and crifis; and makes a decided effort to cut the difeafe fhort at an early period. He is
likewife an enemy to bleeding, purging, vomiting,
and the various evacuations which had been employed
by his predeceffors, attempting to accomplifh his purpofe folely by the means of fweat, and infenfible perfpiration. The fuccefs of his practice was fo great,
that he deems the man unworthy the name of phyfician
who fuffers a fever to exceed the fourth day; a degree
of fuccefs, which all the powers of antimony have not
yet enabled us to boaft of.

The circulation of the blood having been proved
inconteftably about the middle of the laft century,
hopes were reafonably entertained, that the healing
art would be benefited by the difcovery. It does not
however appear that medical men, for fome time at
leaft, either argued more clearly, or practifed more
fuccefsfully. The advocates of the galenical and chemical fchools had gradually approached to each other;
fo that the doctrines and practices of thofe contend-
ing parties were now infenfibly blended together.
Sometimes the one mode of thinking predominated,
fometimes the other; but chemical principles every

Y

where gave fcope to the imagination, which often indulged in the wildeft extravagance of conjecture.
Among the number of thofe conjecturers, who arrived at much eminence and fame, we may reckon
Sylvius de le Bae, who lived in the end of the laft
century, and introduced a confiderable innovation in
the manner of treating fevers. His theories are generally known. They appear to be totally deftitute
of foundation; yet unfortunately are the ground-
work of all his practical indications. His principal
view confifts, in regulating the mixtures of bile and
pancreatic juice. He likewife lays fo great a ftrefs
upon the nature of the occafional caufe, as gives rife
to doubt and ambiguity. Thus he fometimes prefcribes acids, though oftener aromatics, volatiles, and
opiates. But as we poffefs fome cafes, which he appears to have healed, in the Leyden hofpital, with all
his fkill and attention, we are enabled with more
certainty to form a judgment of the particular merits
of his practice. It has not any claim to extraordinary
fuccefs; yet it is evidently innocent of the great
harm which fome later authors have imputed to it.
In fhort, if we except opiates, we may confider the
reft of his remedies as very feeble and ineffectual.

During the time that Sylvius flourifhed in Hol-
land, a new theory of fevers was offered to the public
in England by Dr. Willis, the celebrated author, to
whom we are fo much indebted for bringing into
view the importance of the nervous fyftem, in the
economy of the human frame. It does not however
appear, that this writer's theory ever extended far,
or that it was the caufe of much innovation in practice.

The method of treatment, which was generally
adopted in the fevers of England, at the time when
Sydenham began to ftudy medicine, confifted principally in bleeding, in vomiting with antimonials,
in evacuating the inteftinal canal by means of glyfters
or gentle laxatives; and, in the latter periods of the

difeafe, in attempting to raife fweat by hotter alexi-
pharmics. In the firft conftitution of feafons defcribed by this author, viz. the years 1661,-62,-63,-
and 64, we do not find any material deviation from
this general plan of cure; which was the plan followed by Willis, and other contemporary phyficians.
In the next conftitution, viz. the years 1667, 68 and
part of 69, Sydenham forces himfelf on our obfervation by an attempt to effect a very important innovation. The fever which prevailed during the laft
mentioned years was generally of long duration. It
was ufually accompanied with profufe fweatings, and
often diftinguifhed by petechial eruptions. Cordials,
and hot regimen were fometimes obferved to cut
fhort its courfe abruptly; yet dangerous fymptoms
were ftill more frequently the confequence of this
ftimulating mode of treatment, than a favourable termination. The fagacious Sydenham, inftructed by
repeated experience of the bad effects of this common
method of cure, adopted a contrary one; which he
purfued with boldnefs, and apparently with great fuccefs. It may not be improper to obferve in this
place, that our author is not to be confidered as the
inventor of the antiphlogiftic method of treating fevers.
The ancients, particularly the Arabians, carried the
cooling fyftem ftill farther than the moderns. About
this time however it had fallen into general neglect;
and Sydenham undoubtedly poffeffes the merit of
reftoring it; more perhaps from his own obfervation,
than from a knowledge of what had been done by
his predeceffors. Part of the year 1669, the years
1670,-1671 and 1672, form another conftitution of
feafons, according to this author's arrangement of
difeafes. The epidemic affumed a different appearance from the former. It was chiefly diftinguifhed
by fymptoms of dyfenteric affection. Our author,
however, ftill adhered to the outlines of the antiphlogiftic plan; and treated the difeafe fuccefsfully with

bleeding, and the repeated ufe of laxatives. The method of treatment, which he adopted, admits of a remark. In the former epidemic, the profufe fweatings were checked; in the prefent, the inteftinal evacuations were encouraged; in one cafe he appeared to promote, in the other to thwart the intentions of nature; practices fo oppofite that we cannot eafily reconcile them. The next conftitution, viz. the years 1673,-74 and 75 difcovered a fever with a new train of fymptoms, and in Sydenham's opinion of a very different race. It was principally diftinguifhed by pleuretic and rheumatic affections, by coma and ftupor. The general antiphlogiftic practice was ftill perfifted in; and the whole of the cure was trufted to difcretional bleeding, bliftering the back part of the head and neck, with the repeated employment of glyfters. The hotter diaphoretics were cautioufly avoided. In the year 1684, this diligent obferver imagined he difcovered the appearance of a fever of a perfectly new and unknown kind; a fever accompanied with more or lefs derangement of intellect, and many other fymptoms of nervous affection; the fpecies of difeafe, perhaps, which nofologifts have diftinguifhed by the name of Typhus. But though this fpecies of fever was fuppofed by our author, to be extremely different in its nature from any that he had yet feen, we do not however perceive, that this idea fuggefted to him any material difference in the mode of treatment.

From the fhort view which has been given of Sydenham's practice in fevers, it is eafy to perceive the rife and progrefs of the method of cure which he adopted. Antiphlogiftic proceffes were carried to a greater length by the ancients, than the moderns have yet dared to rifk. But there is little reafon to fuppofe, that Sydenham owed the ideas of the alterations which he introduced to information from preceding writers. His practice bears every where authentic

marks of having arifen from his own obfervation.---
The moft common termination of fevers, is by fweating or increafed perfpiration; a fault obferved by
Van Helmont, and which furnifhed that author with
the idea of profecuting the cure of the difeafe wholly
on this plan. The practice feems to have been early
adopted in many parts of Europe; and it even continued in general reputation in England, at the time
that Sydenham began his medical ftudies. Sweating
undoubtedly is often beneficial, and may be confidered, upon the whole, as the moft certain means of
exterminating the caufe of fevers; yet bad effects often refulted from it---then probably more owing to
the manner in which it was conducted, than to the
real hurtfulnefs of the thing itfelf, viewed in the light
of a general remedy. Sydenham, who does not appear to have difcriminated between the actual effects
of fweating and the effects of the manner of exciting
it, condemns the practice in general terms, and
paffes to an oppofite method of treatment with a good
deal of boldnefs. It has ever unfortunately been the
fate of phyfic, like every other conjectural art, to
pafs from one extreme to its oppofite by large ftrides;
and thus, even the fagacious Sydenham, who had feen
the bad effects of treating remedies in fevers with
much of the inflammatory diathefis, was induced to
employ antiphlogiftic procefies in thofe fpecies of difeafe, which we fhould be difpofed to believe do not
well admit of them. The new, or nervous fever,
in the opinion of the practitioners of the prefent age,
could not well bear the plentiful evacuations prefcribed
by this author; at leaft, we may fafely affirm, that
fuch evacuations are not by any means neceffary.---
But I fhall difmifs this fubject with obferving, that the
practice of Sydenham, if we except the article of
bleeding, can only be confidered as feeble, and as often infignificant. His remedies fometimes, perhaps,
obviate the fatal tendency of fymptoms; but are not

capable of having any decided effects on the natural
courfe of the difeafe. I may likewife add, that his
practice is directly at war with the principle of his
theory. If fever is confidered an effort of nature to
exterminate fomething hurtful from the conftitution,
bleeding and thofe evacuations, which diminifh the
powers of life, are not the proper means of effecting
this purpofe. But the truth is, the practice of Sydenham was his own; his theory was that of the
times in which he lived, formed from a mixture of
the doctrines of Van Helmont, Campanella and Dr.
Willis.

It may not be fufpected, perhaps, from the remarks
which I have made on the practice of Sydenham in
fevers, that I do not confider him as the author of fo
much effential improvement, as has been generally
imagined. I muft however acknowledge, that he
deferves the higheft praife for the accurate and well
difcriminated hiftory of acute difeafes, which he has
left us. The defcriptions are complete, and the circumftances fo peculiarly chofen, that the difeafe itfelf
is actually before the eyes of the reader. Thefe are
the great, and as yet the unrivalled excellencies of
Sydenham; but in admitting fuch effential differences
in the caufe of epidemics as he has done, he neceffarily leads us to embarraffment, and often leaves the
practitioner in a ftate of uncertainty. The difeafe
defcribed by Sydenham, in the various conftitutions
of feafons between the years 1661 and 1685, fhews
external marks of confiderable diverfity; yet I muft
confefs, that I fee but little reafon for fuppofing, that
thefe appearances arife from caufes which are totally
and fundamently diftinct. The fever of Sydenham,
in all its forms, is in fact the common endemic of
England. Circumftances however often arofe then,
and ftill arife, which modify the general caufe in fuch
a manner, that the difeafe appears at one time with
fymptoms of inflammatory diathefis, at another with

fymptoms of nervous affection, and at another, with a general difpofition to affections of particular organs. Thefe modifying caufes, which are more general or particular, more obvious or obfcure, often continue for a certain train of feafons, and influence very materially the character of the reigning epidemic. The general caufe of the fever is in reality one and the fame, yet I muft alfo acknowledge, that the modifications are evidently many and various, and often very remotely different from each other.

Chemical principles for fome time paft, had the principal fhare in enabling medical writers to account for the phenomena in fevers ; but about the end of laft century, the mechanical philofophy was again revived, and being incorporated with the doctrines of the chemifts, the laws, and various derangements of the human frame, were then explained on the principles of hydraulics, or chemical mixture. The authors who adopted this mode of reafoning were numerous, and fome of them were of great eminence; but at prefent I fhall only take notice of one of the greateft of them, the celebrated Boerhaave, who formed a fyftem, which was confidered as the moft perfect that had hitherto been offered to the public. The doctrines of this author acquired uncommon fame. They foon extended over all Europe, and, indeed, ftill prevail in the greateft part of it. But though Boerhaave has prefented us with a methodical explanation of the phenomena in fevers; and has detailed the method of cure with clearnefs and precifion ; yet we do not find, that he has furnifhed much that is new and original in practice. He is every where cautious, and in moft inftances judicious; though he has committed a principal error in forming indications of cure, from a fuppofition of lentor and vifcidity ; a caufe the very exiftence of which we have every reafon to doubt.

During the time that Boerhaave flourifhed in Hol-

land, indeed before this author had arrived at much
reputation, Profeſſor Stahl, at Halle in Saxony, pro-
poſed ſome new opinions, which acquired conſiderable
fame, and which have been conſidered, in ſome manner, as forming a peculiar ſyſtem. The leading
principle of this author, as is confeſſed by all, admits
only of a feeble and inactive practice. I might even
add, that it frequently leads to a pernicious one.
Thoſe tumults, or ſufferings, which paſs by the name
of the efforts of nature, are extremely deceitful;
and have obviouſly, in many inſtances, a deſtructive
tendency. I mentioned before that they are truſted
with danger; yet Stahl, proceeding on this principle, boaſts extraordinary ſucceſs in the cure of
the petechial fever, which prevailed in moſt parts of
Saxony towards the end of laſt century.

In a review of thoſe authors, who have written
on febrile diſeaſes, it would be unjuſt to omit mentioning Hoffman, contemporary with Stahl, and profeſſor in the ſame univerſity. The actual alterations
which this author has introduced into the cure of
fevers, are not perhaps very great in themſelves;
yet his important diſcoveries, in regard to its theory,
entitle him to great conſideration. The moſt of the
remedies, which he employed, are found in the writings
of his predeceſſors, or contemporaries; yet they were
not, perhaps, always preſcribed by them with the
ſame intentions. The theory of Hoffman opens a
road for the trial of antiſpaſmodics, merely on the
footing of antiſpaſmodics; a claſs of remedies of
much importance in the cure of febrile diſorders. In
practice, Hoffman is more decided than Stahl; and
his views, perhaps are more extenſive than thoſe of
Boerhaave. He is likewiſe uncommonly candid; and
has furniſhed us with a great variety of hiſtories,
which ſerve in many caſes to illuſtrate the nature of
the diſeaſe.

The antiphlogiſtic method of treating fevers, the

ground-work of which was laid by Sydenham, and improved by Boerhaave, prevailed in moſt parts of Europe, without material alteration, till near the preſent times. Bliſtering with cantharides, which had been employed with caution, and which was even ſuſpected of deleterious effects by many, was introduced into practice in the end of the ſixteenth century, aad about the beginning of the preſent began to be, as employed, a common remedy in many ſpecies of fever : its good effects were often obvious, and, according to the prevailing mode of reaſoning, were ſuppoſed to ariſe from a quality which cantharides were believed to poſſeſs, *of attenuating* the blood. This mode of operation is no longer admitted; but the remedy ſtill retains its credit. Few people pretend that bliſters are poſſeſſed of ſpecific powers in ſhortening the courſe of fevers ; yet every one allows, that they obviate many ſymptoms of dangerous tendency, and that they often diſpoſe the diſeaſe to aſſume it proper form. In fevers, accompanied with local affection, their beneficial effects are univerſally acknowledged ; and, even in many caſes of general irritability, they often produce very fortunate changes. But I muſt obſerve, with regard to this, that much depends on managment, and the mode of application. In local affections the local application is moſt effectual ; in caſes accompanied with much general irritability, the back part of the head and neck ought, perhaps, to be preferred to others. I have thus frequently ſeen in fevers, where there was much general irritability, that bliſters applied to the extremities evidently aggravated the diſeaſe; while I have alſo obſerved, that they as certainly diminiſhed the hardneſs and frequency of the pulſe, and diſpoſed the patient to reſt, where they were applied to the back part of the head and neck. There is another remedy that I ſhall take notice of before leaving this ſubject, which poſſeſſes ſtill higher reputation than bliſters.

Antimonial preparations have been employed occafionally in fevers for many years paft; but they did
not gain eftablifhed credit in this country, till within
thefe thirty years. The difcovery of the famous
powder of Dr. James appears to have been the caufe
of a confiderable innovation, in the manner of treating febrile difeafes. The practice of Boerhaave did
not go farther than to obviate fymptoms of fatal
tendency; it left the difeafe to purfue its own courfe.
Dr. James affumed a bolder ground, and promifed to
cut fhort the fever abruptly by means of his powder.
There are many who ftill tread in his footfteps; I
acknowledge, as I have hinted before, that their attempts may be often fuccefsful in the early ftages of
the illnefs, or often ufeful towards a critical period.
I cannot however believe, that this powder, or any
preparation of antimony with which we are yet acquainted, poffeffes the power of abruptly terminating
a fever wherever it is employed; at leaft, to effect
this requires a management of which I confefs myfelf
ignorant. The effects which Dr. James promifed
from his powder, others have attempted to obtain
from emetic tartar; but I have reafon to think with
inferior fuccefs.

The wonderful power, which the Peruvian bark is
obferved to poffefs, in fufpending the courfe of intermittents, has led the practitioners of the prefent times
to employ it, with the fame views, in fevers of various
denominations. But after what I have faid of the
uncertainty of its effects in checking the courfe of
the remitting fever of Jamaica, it will be needlefs to
repeat here, that I do not expect to find it of much
efficacy, in fhortening fevers of a more continued
kind. I muft, however confefs, that, even in many
of thefe, it is a remedy of great value. It fupports,
in a very eminent degree, the tone and vigour of the
powers of life.

Opium has been prefcribed occafionally in fevers

for a long time paft; but it is only of late years, that it has been recommended, as a general remedy in fome particular fpecies of this difeafe. The practitioners of the Weft Indies, prefcribe opium with more freedom, than is generally done in England. It is frequently employed to mitigate fymptoms; and in fome fituations which were very alarming, I have given it in very large quantity with unexpected good effects. In the flow fevers of this country I have frequently had recourfe to it; and, combined with antimonials and camphire, have found it to be a remedy, of great value. Opium in general was more cordial than wine. In cafes of defpondence and diftrefs it gave a confidence to the mind, and imparted a pleafureablenefs to the fenfations above all other remedies. In fhort, it appeared often, not only to be inftrumental in conducting the difeafe to a favourable termination, but it enabled the patient to pafs through it with comfort to himfelf.

I have mentioned in the preceding pages, the moft eminent of thofe authors, who have written on the cure of fevers; giving at the fame time fuch extracts from their works, that the reader, who has not the opportunity of confulting the originals, may be enabled to form fome idea of the fucceffive changes, the improvements, and oftener perhaps the corruptions, which have arifen in the method of treating febrile difeafes, from the earlieft records of the art to the prefent times. The apparent changes are more numerous than the real ones; while the moft oppofite modes of treatment do not often appear to have much perceptible effect on the event. The cure of fever has been hitherto purfued on two general and oppofite views, viz. on the idea of exciting the powers of life, by means of heating and ftimulating remedies; or of diminifhing the reaction of the fyftem by evacuations and other antiphlogiftic proceffes. The above extremes of thofe directly oppofite modes of treatment have approached gradually to each other, or been varioufly combined by different practitioners. It cannot however fail of appearing ftrange to a perfon, who views the fcience of medicine in a philofophical light, to hear one fet of men afferting that the proper cure of fever confifts in exciting the powers of life, or in enabling nature to expel the difcafe by force; while another, with no lefs confidence, maintains that the plan of moderating or diminifhing increafed action is that which ought alone to be purfued. From fuch contradictory affertions we cannot eafily avoid concluding, either that the moft oppofite means produce the fame effect, or that nature has a prefcribed mode of proceeding in fevers, which ordinary medical affiftance is not powerful enough to controul. There are many eminent practitioners, who have been confcious of this truth. The candid Sydenham himfelf acknowledges, that thofe, whom he treated with all his fkill and attention, and who poffeffed all the comforts that affluence could afford, did not often fare better than the poor, who were only fparingly furnifhed with neceffaries, and who met with little affiftance from medicine. I have myfelf feen many examples of the fame kind. Sometimes I purfued the ufual methods of cure with care and perfeverance; fometimes I left the bufinefs almoft entirely to nature, and I cannot fay, that the difference of the event gives me much caufe to be vain. But though I may appear to be fceptical with regard to the effects of common practice, I ftill cannot help being of opinion, that we may arrive at a high degree of perfection in the management of febrile difcafes. So fanguine, indeed, are my expectations, that I cannot eafily forgive myfelf, when the event of this difeafe happens to be unfortunate. The remitting fever of Jamaica is not a difeafe by any means devoid of danger; yet I fhould not be fatisfied with myfelf, from the view which I now have of the fubject, if I loft one patient in fifty. I own indeed that this is a degree of fuccefs, which neither I, nor perhaps any other man has yet attained. I muft however add, that I have not always had the liberty of doing what I wifhed to do; neither have I always dared to venture upon what I judged not proper to be done. The prejudices of patients in fome cafes, and the idea of refponfibility in others, confine us to the beaten track, though we may be confcious in ourfelves that it never can lead us to our object. If thefe obftacles were removed, a man who will act with decifion, may promife almoft any degree of fuccefs in the remitting fever of the Weft-Indies, in conftitutions which are free from habitual complaints.

The conftant fluctuation which has hitherto prevailed in the opinions of phyficians concerning the caufes of fever, and in their practices with regard to its cure, oblige us to think doubtfully of the real progrefs of the healing art. Hippocrates was allowed to have practifed with more fuccefs than his predeceffors. Afclepiades was believed by many to have been ftill more fortunate than Hippocrates ; yet the road which he purfued was totally different. Galen, who reviewed and improved the fyftem of the Coan fage, rofe to great eminence, and marked out the path of medical practice for many centuries. The doctrines of Paracelfus fhook his authority ; and thefe in their turn gave way to newer modes of thinking. In this manner there have been fuch perpetual revolutions in the modes of treating febrile difeafes, that we can fcarcely avoid concluding, that little or nothing of the matter is yet known with certainty. Medical writers have wandered from conjecture to conjecture, for more than two thoufand years ; and we do not yet perceive any profpect of thefe conjectures being nearer to an end.

Z

# APPENDIX.

## CONTAINING SOME HINTS WITH REGARD TO THE MEANS OF PRESERVING THE HEALTH OF SOLDIERS SERVING IN HOT CLIMATES.

HAVING treated pretty fully of the remitting fever of Jamaica, and intermitting fever of America, I shall now offer a few thoughts on the various means of preserving the health of soldiers in warm climates; taking the liberty at the same time to suggest some ideas, which might perhaps be usefully attended to by those who superintend the medical establishments of the army.

The climate of the West-Indies has been fatal to the European constitution, even since its first discovery by Columbus. To the armies and navies of England it has been particularly destructive. The sad fate of the troops who went on the expedition to Carthagena will be long remembered; neither will the loss sustained at the Havannah, Martinique and Gaudaloupe soon be forgotten; while the destruction, occasioned by the effects of climate at St. Lucia, St. Juan, and even in Jamaica, during the late war, is still fresh in our memories. As it appears from a comparative view of the mortality of the troops employed in these different services, that we have profited but little by the experience of our former misfortunes, it might probably be supposed, that the great sickness, observed on these occasions, has actually arisen from the irremediable effects of climate, or unavoidable hardships of service in hot countries; but there is reason to believe that this is not wholly the case. I will venture to assert, nor should I expect to meet with difficulty in proving, that much of it has proceeded from the inexperience or inattention of thofe who conducted the expeditions, or from fuch errors in the medical departments as might have been eafily obviated. It is fuperfluous to obferve, that the health of the foldier is an object of principal importance in enfuring the fucceffes of war. We have many inftances of expeditions apparently well concerted, which have failed from the exceffive ficknefs of the troops: and too many proofs of this ficknefs proceeding from a neglect of fuch precautions, as might have contributed to the prefervation of health. I have accuftomed myfelf to look at this fu ject for more than fifteen years. I have turned it often in my mind, and cannot difcover that much judicious attention has yet been paid to it. We cannot often perceive that health has been an object of confideration, in fixing the permanent ftations of troops; or that it has been much regarded in choofing encampments in the field. Exercifes, which might inure the body to hardfhips, have not been fufficiently enforced; and fuch forts of diet, and fuch modes of life, as might obviate the danger of difeafes, have been little attended to; while the beft regulations for a fpeedy and decifive plan of cure do not appear to have been adopted. I fhall be obliged, in tracing this fubject, to advance fome ideas which are contrary to the opinions of fome celebrated authors, which combat popular prejudices, or which interfere with the views of interefted men. I may be reckoned prefumptuous perhaps in cenfuring freely; but I am confcious that I do not advance any thing which has not truth for its foundation.

It has frequently been the practice, in times of war, to fend new raifed regiments to ferve in the iflands of the Weft-Indies; and though the injudicioufnefs of the practice has long been difcovered, it does not yet appear to be difcontinued. During the late war there were feveral corps fent out to thofe countries newly recruited, the confequence of which was, that though not a man died by the fword; yet in the fhort fpace of two years, there fcarcely was a foldier left. A great part of this dreadful mortality undoubtedly arofe from the climate; yet fome fhare of it feems likewife to have proceeded from the particular circumftances of raw undifciplined troops.— Men newly enlifted in England, are generally of grofs and full habits, and too often accuftomed to irregular and diffipated modes of life. Under fuch circumftances, a fudden tranfition to a hotter air, joined with full meals, and the habitual indolence of a paffage at fea, cannot fail to produce a plethoric ftate of the body, which is often rendered dangerous by the incautious ufe of ftrong liquors, or the ordinary exertions required in performing military exercifes, under the influence of a powerful fun. I do not pretend to infinuate that thofe are the caufes of remitting fever, but I am very fenfible at the fame time that they are caufes which occafionally aggravate its danger, and which even fometimes accelerate its appearance. In foldiers who have been inured to a military life, fuch change of climate operates with diminifhed effect. The bulk of the fluids is perhaps diminifhed by a continuance of lefs full living; while the tone and elafticity of the moving powers are increafed by habits of exercife and exertion.— The difpofition to commit exceffes is likewife reprefled by the rigour of difcipline; and the mind acquires a philofophical firmnefs from long fervice, which not only contributes to the prefervation of health, but which enables the individual to fuftain with fortitude the attack of difeafes.

In paffing from a cold to a hot climate, the firft thing that occurs to be confidered, is the effect produced by the fimple increafe of heat on the human frame. Expanfion of the fluids, and confequent fulnefs of the veffels is conftantly obferved to take place

from fuch a change, frequently however accompanied
with diminifhed energy of the moving powers, particularly where heat is combined with dampnefs of
the air. To obviate therefore this natural effect of
heat is the firft general object to be attended to, in
tranfporting troops to the tropical climates. The
Englifh, from the conftitution of their bodies, and
ftill more perhaps from their manner of living, fuffer
more from thofe fudden changes than fome other
European nations. The French and Spaniards are
not only lefs grofs conftitutionally, but eat likewife
lefs animal food, and drink their liquors greatly more
diluted, than the natives of England. They do not
probably owe more to medical affiftance than the
Englifh ; yet they are known to efcape better from
dangerous difeafes ; and their fafety I might add has
been remarked to bear fome proportion to the different degrees of abftemioufnefs, which they are known
to obferve. An idea prevails with the generality of
people, who vifit warmer or more unhealthy climates,
that it is neceffary to eat and drink freely, as a fecurity againft the attacks of endemic fevers ; but a
very narrow obfervation will ferve to fhew, that good
living, as it is called, has no fuch effects ; and we
may even foon perceive, unlefs blended by long
eftablifhed prejudices which flatter our appetites,
that it actually is attended with pernicious confequences. The moft abftemious, fo far as I have obferved, efcaped the beft, not only from the attacks,
but particularly from the danger of difeafes. With
regard to the diet of a foldier, ferving in a hot climate, I fhould be difpofed to believe, that one fpare
meal of animal food would be perfectly fufficient in
twenty-four; and if it were eafy to alter eftablifhed
cuftoms, it would be moft proper, perhaps, that it
were made in the cool of the evening. Coffee, or tea
for breakfaft might likewife be fubftituted with advantage in place of the ordinary allowance of rum :

but this I muft confefs would be a dangerous experiment, Our foldiers have been fo long accuftomed
to confider this gratuitous allowance of rum as their
right, that no man could anfwer for the confequences
of with-holding it. The practice certainly is pernicious, and the man, who firft introduced it into the
army, did no good fervice to his country. I do not
deny that a judicious ufe of fpirits might be of benefit
occafionally: neither do I pretend to fay, that, even
the hardeft drinking can be confidered as a general
caufe of fevers; but it would not be difficult to produce evidence, that hard drinking aggravates the
violence, and increafes the danger of the difcafe, when
it happens to take place ; while I cannot perceive
much reafon for concluding, that the ufe of fpirituous
liquors has ever been productive of general good to
the army, particularly in warm climates. But as I
have juft mentioned, that fpirituous liquors have little
claim to be confidered among the number of thofe
things, which contribute to the prefervation of health :
fo I may add, with perfect confidence, that the allowance of rum granted to foldiers, has done much harm
by ruining difcipline, and good behaviour. If it is
with-held for one day, difcontent immediately begins
to fhew itfelf among the men. If with-held for any
length of time, complaints fometimes rife to a ftate
of mutiny, and defertions become numerous. But
befides this, that foldiers feldom perform extra-duty
with alacrity, unlefs they are bribed with a double
allowance of liquor. A double allowance, drank un-
diluted, as is generally the cafe, is frequently fufficient to produce fome degree of intoxication. I need
not mention the difafters to which an intoxicated
army is expofed. Difafters of a very ferious nature
have actually happened from this caufe, and they
might have happened oftener had the enemy been
always vigilant, and bold enough to have feized the
opportunity.

A deal might be faid on the fubject of abftemioufnefs. Moderation both in eaing and drinking is effentially neceffary to the health of troops newly arrived in hot climates; but a truth fo obvious need
not be enforced by many arguments. The example
of the French and Spaniards afford a very convincing one. It is known to every medical perfon, that
the fevers of hot climates are generally moft dangerous in full and plethoric habits. It ought to be an
object of attention therefore to obviate this caufe of
mortality, by means of fpare living, and the cautious
ufe of ftimulating liquors : but foldiers have little felf
command, and feldom refift the gratification of their
appetites. Hence it becomes the duty of their officers to enforce their compliance with what is proper,
and to preclude them, as much as is poffible, from the
means of obtaining what is pernicious; but this requires great vigilance and attention, and often great
feverity. It is not enough that foldiers are obliged
to eat in meffes. The officers ought daily to infpect
their meals, and inflict penalties where they obferve
tranfgreffions. And further, as it is a matter of much
importance to preferve troops in a ftate of health fit
for action, and as the courfe of fevers is often uncommonly rapid in the Weft-Indies, it would be proper, perhaps, that the furgeon reviewed the men daily.
The diftant approach of the difeafe would be frequently difcovered by this means, and the danger of
it might probably be fometimes averted by timely affiftance. Before men appear in the fick-reports, the
fever is often confiderably advanced in its progrefs.

Befides the alterations which might be made in the
diet of troops, on their arrival in hot climates, fome
changes in the mode of cloathing might, perhaps, be
likewife adopted for the fake of eafe and convenience,
if not for purpofes of real ufe and economy. Round
white hats would be the moft proper covering for the
head; and dowlas might be fubftituted with advan-

tage in room of the thick cloth, of which the coats of foldiers are ufually made. There can be no grounds for fuppofing, that a foldier will not fight as well in dowlas as in fcarlet; and there is certain proof that he will perform duties, which require exertion, with greater fafety and effect, as the nature of his cloathing will preferve him cooler by fome degrees. But though fuch alterations may be hinted, there is little room to believe that they will be attended to. In the prefent rage for military fhew, it will be a difficult tafk to convince men to lay afide an uniform, which adds fo much to the brilliancy of the appearance. Much ftrefs feems at prefent to be laid upon the drefs of the foldier, and I do not pretend to argue, that it is a matter of perfect indifference. It has certainly very often had vifible effects upon the enemy; but thefe effects have oftener proceeded from a knowledge of the character of the troops who wore it, than from any thing formidable in the uniform itfelf. But to leave this fubject of drefs, I fhall only obferve, that a flannel or cotten wrapper would be more ufeful to a foldier, ferving in the Weft-Indies, than a blanket; and perhaps the expence of it would not' be much greater. It would ferve for his covering in the night, and would fecure him againft the effects of cold, where occafions obliged him to go out.

I fhall endeavour in the next place to point out fome of thofe benefits, which may be derived to health, from habits of daily exercife. This is an object of the greateft importance, but unfortunately it is an cbject very little attended to in the Britifh army. It appears, indeed, to be little regarded in moft of the armies of modern Europe. I fhould incur a charge of prefumption, perhaps of ignorance, did I attempt to point out the exercifes which are the moft proper for the forming of foldiers. Thofe only which contribute to the prefervation of health, belong to this place. I may however remark, that the effential part of the

art of difciplining troops, confifts in imparting fenti-
ments of heroifm and virtue to the minds of the men, in
improving the exertions of their limbs, and in acquiring knowledge of the correfpondence of their exe-
tions when called into action. If I durft take fo great
a liberty, I fhould be inclined to fay, that our ordinary
exercifes are flat and infipid in their nature; that they
occafi n no exertions, and excite no emulation: they
neither improve the active powers of the body, nor
inure the foldier to bear fatigue and hardfhip. The
Romans, who owed more to the difciplme of their
armies than any nation on earth, were extremely ri-
gorous and perfevering in their exercifes. They practifed their foldiers in every fpecies of fervice that
might occur; fo that nothing at any time happened
with which they were unacquainted. Actual war
was in reality a time of relaxation and amufement to
the foldiers of this warlike people, who appear to have
been trained for the fervice of the field, as horfes are
for hunting or the courfe. The Romans were not
only fenfible of the advantages which thofe habits of
exercife procured them in action; but had al o the
penetration to difcover, that they were eminently ferviceable in the prefervation of health. The words
of Vegetius are remarkable. " Rei militari, peric̣i,
plus quotidiana armorum exercilia ad fanitatem mili-
tum putaverunt prodeffe, quam medicos." I made the
fame remark during the time that I attended a regiment in America, without knowing that it was fupported by fo great authority. I obferved, when the
men were in the field, fometimes even complaining of
hardfhip and fatigue, that few were reported in the
lift of the fick: when removed to quarters, or encamped for any length of time in one place, the hofpital was obferved to fill rapidly. This obfervation
was uniformly verified, as often as the experiment
was repeated.

An idea has been long entertained, that the Euro-

ropean conflitution cannot bear hard labour in the
fun, or perform military exercifes with fafety, in the
hot climates of the Weft-Indies. Hence a plan has
been fuggefted, and in fome degree I believe adopted,.
that regiments ferving in thofe countries, be furnifhed'
with people of colour to do the drudgery of the foldiers. But this appears to be an innovation which
ought to be admitted with extreme caution. It will
evidently ferve to increafe floth and idlenefs, and unlefs the perfons of colour can perform the military
duty in the field, their fervices will go but a fhort
way in preferving the health of the troops. A foldier, notwithftanding he may have received the King's
pay for twenty years or more, remains in fome degree a tyro till his body has been inured to fatigue,
and prepared to bear without danger the effects of
the climate, in which he may be deftined to ferve.
This is a part of the military difcipline, indeed, no
lefs neceffary than a knowledge of the ufe of arms ;
and though it is a part of it, difficult to be accomplifhed, there is ftill room to believe, that it may be
effected, even in the fo much dreaded climate of Jamaica. It is a common opinion, that the fatigues
of an active campaign in the Weft-Indies, would be
fatal to the health of the troops ; but the opinion has
been affumed without fair trial. The exertions of
a fingle day have often been hurtful. This was frequently the cafe in America, where the foldiers had
remained for fome time in a ftate of reft ; but bad
effects from the greateft exertions, in the hotteft
weather of fummer, were extremely rare in that
country, after the campaign had been continued for a.
few days. But that I may not feem to reft an opinion of fo great importance on a bare analogy, I fhall
beg leave to obferve, that young European planters
undergo greater fatigues, and remain daily expofed
for a longer time to the heat of the fun, than would'
fall to the lot of foldiers in the actual fervice of the

field.  I might likewife further  confirm the opinion, that an Englifhman is capable of fuftaining fatigue in the Weft-Indies, equally  well  with the African, or the  native  of  the  iflands, by  mentioning  a journey which I once performed myfelf.  I lived about four years in Jamaica, during  the  greateft part of which I believed  that death, or dangerous  ficknefs, would be the confequence of walking any  diftance on  foot ; but I afterwards  learnt that  this apprehenfion was vain.   I left Savanna la Mar in the year 1778, with the  defign  of  going to America;  but  having em-barked  in  a hurry,  and  forgot  a  material  piece of bufinefs, I found a neceffity of being  put afhore, after having been two or  three days at fea.   I was landed at Port Morant, in  St.  Thomas's  in the Eaft, from which I went to Kingfton by water, where  learning that there  was a veffel at Lucca, in the Weftern ex-tremity  of  the  ifland, nearly  ready to fail for New-York, I fet out directly,  that I might  not lofe the opportunity of a paffage.   My finances  not being in a condition to furnifh horfes, I left Kingfton on foot, about  twelve  o'clock,  and  accomplifhed a journey before it was dark of eighteen miles.   I did not find I was  materially fatigued  and ftill perfifting in my refolution,  travelled  a  hundred  miles  more in the fpace of the three following days.   It may not be im-proper  to remark,  that I carried baggage with me, equal in weight to the common knapfack of a foldier. I do not know that fo great a journey was ever per-formed on foot by an European, in any of the iflands of  the Weft-Indies ;  not fo much I am  convinced from  inability, as  from  idea that fuch exertions are dangerous.   But as  it appears from the  above fact, that  the  European conftitution is capable of fuftain-ing  common military fatigues in the climate of Ja-maica ;  fo  I  may add  that it ought to be a principal object of military difcipline, that foldiers be practifed with  frequent  marching,  and  the  performance of other exercifes of exertion, if it is actually meant that they fhould be ufeful in times of war. The fate of battles, I might obferve, depends oftener on rapid movements, in which the activity of the limbs is concerned, than on the expert handling of arms, which is acquired by the practice of the manual. I obferved formerly, that abftemioufnefs and temperance were among the beft means of preferving health, or obviating the danger of the difeafes to which troops are liable on their firft arrival in hot climates; but the rules of temperance are little regarded by Englifh foldiers at any time, and almoft conftantly tranfgreffed wherever extraordinary labour is required of them. To fuch caufes of excefs, joined with the great heat of the fun, we may perhaps impute many of the bad effects of marching, or of moderate fatigue in the Weft-Indies. In the journey which I have juft now mentioned, I probably owe my efcape from fick-nefs to temperance and fpare living. I breakfafted on tea about ten in the morning, and made a meal of bread and fallad, after I had taken up my lodging for the night. If I had occafion to drink through the day, water or lemonade was my beverage. In the year 1782, I walked between Edinburgh and London in eleven days and a half; and invariably obferved, that I performed my journey with greater eafe and pleafure, where I drank water, and only breakfafted and fupped, than when I made three meals a day, and drank wine, ale, or porter. In the fol-lowing fummer I carried the experiment farther. During the months of July and Auguft, I travelled in fome of the hotteft provinces of France. I ge-nerally walked from twenty-five to thirty miles a day, in a degree of heat lefs fupportable than the common heat of Jamaica, without fuffering any material in-convenience. I breakfafted about ten o'clock on tea, coffee or fyrup of vinegar, made a flender meal of animal food in the evening, with a great proportion

A a

of fallad and vegetables; but never drank the weakeft wines without dilution. The great refrefhment which I found from fyrup of vinegar and water, convinces me, that the Romans had good caufe for making vinegar fuch an effential article among the provifions of their armies.—The ftate of luxury and our depraved appetites, unfortunately do not fuffer it to be adopted by the Englifh. I ought perhaps to make an apology to the reader for introducing my own experience on the prefent occafion : but I muft add, that I have only done it, becaufe it enables me to fpeak from conviction, that an Englifh foldier may be rendered capable of going through the fevereft military fervice in the hotteft iflands of the Weft-Indies, and that temperance will be one of the beft means of enabling him to perform his duty with fafety and effect.

I mentioned before, that the military exercife of the Englifh army is ill calculated to excite a fpirit of emulation among the men. It is in fact confidered only as a piece of drudgery, in which there are few who have any ambition to excel. It has little effect in improving the activity of the limbs, or hardening the conftitution of the body; fo that it may better fuftain hardfhip and fatigue. But feeble as its effects are in the view of increafing exertion, or preferving health, it is generally almoft intirely difcontinued when troops arrive in hot climates ; a practice, which has arifen from a fuperficial and miftaken view of the fubject. Sloth and indolence are the bane of a foldier in every climate ; exercife and action are the greateft prefervatives of difcipline and of health. It would be reckoned prefumption in me, and it does not belong to this place to point out thofe exercifes which might be proper for the forming of foldiers. But every one knows that walking, running, wreftling, leaping, fencing and fwimming, are often called into actual ufe in the practice of war. Thefe are fuch exercifes likewife as excite emulation, and are practifed with pleafure by the individual. They harden the body, increafe the power of the limbs, and by furnifhing the officer with a view of the different degrees of activity, may often enable him to place his men in the ranks, according to the uniformity of their exertions ; a more ufeful mode of arrangement in time of action, than uniformity of exterior form. I may add in this place, that fea-bathing will be extremely ufeful in moft cafes, in increafing the vigour and preferving the health of foldiers ferving in warm climates. There no doubt will occur many cafes, in which it is improper ; but in general it may be employed with great benefit. I chiefly impute it to this caufe, that I did not experience a fingle day's indifpofition, during the four years that I lived in Jamaica.

It has been known for many ages, that the caufe of intermitting and remitting fevers, the moft formidable difeafes of hot climates, owes its origin to exhalations from fwampy and moift grounds. It often happens likewife, that thofe low and fwampy grounds are the moft acceffible parts of a coaft, or that towns and fettlements have been placed near them—to attack or defend which falls to the lot of the foldier. It not being therefore in the power of a military commander to remove the natural difadvantages, which I have mentioned ; it is only in his power to fhew his judgment and attention, by applying the beft remedies to obviate their effects. It is certainly an object of the utmoft confequence to preferve troops in a ftate of health fit for action : and no perfon will deny, that every care ought to be employed in choofing the beft fituations for quarters, or even temporary encampments, that the nature of the duty will permit. We learn from experience that fevers are little known in rough and hilly countries, where water flows with a rapid courfe ; while we likewife

A a 2

know, that they are common in low and champaign
countries, where water ftagnates, or has only a
fluggifh motion: independent of which, thofe fituations which are in the neighbourhood of fwamps, or
near the oozy banks of large rivers, have always been
obferved to be particularly liable to fuch difeafes. If
therefore the circumftances of the fervice do not for-
bid, no room can be left to doubt about the propriety
of ftationing troops in the mountanous or hilly parts
of a country; while I may likewife add, that where
neceffity confines them to the plain, the fea fhore
will in general be found to be the moft eligible. But-
befides the above general character of local fituations
there are likewife fome fubordinate circumftances,
which deferve to be particularly attended to in choofing
the ground of encampments. It is very commonly
believed that high and elevated fituations are the moft
uniformly proper for this purpofe; but this in fact is
not, by any means, a general rule. A high and dry
fituation does not contain any thing hurtful in itfelf;
but it is more than others expofed to the effluvia
which are carried from a diftance. It is the peculiar
nature of exhalations to afcend as they proceed from
their fource; in confirmation of which truth I have
had feveral opportunities of witneffing, that this caufe
of difeafe was carried to rifing grounds in a ftate of
great activity; while it apparently paffed over the
plain or vallies which lay contiguous, without producing any material effects. From the knowledge
of this fact we are furnifhed with this obvious remark, that it will be proper to interpofe woods or
rifing grounds to the progrefs of thofe noxious vapours; or where fuch natural advantages do not exift,
it might be ferviceable to burn a chain of fires in a
temporary encampment, or even to raife a parapet
wall to over top the barracks, where neceffity requires a more permanent ftation.——It would be a
matter of utility, could we determine with any cer-

tainty to what diftance from its fource, the noxious effluvia extend; but this is a queftion which we cannot hope to afcertain very exactly. It is not uniformly the fame in all fituations, depending on the concentrated ftate of the exhalation at its fource, the obftacles it meets with in its progrefs, and the nature of the ground over which it paffes, or to which it is directed. I have known its influence very remarkable at the diftance of a mile and a half, on the top of a hill of very confiderable elevation.

The conveniences of trade have often tempted colonifts to place their towns on the banks of rivers. without regard to the healthfulnefs of the fituation.— The choice of fuch fpots, injudicious as it evidently is, has been greatly approved of, and warmly recommended as preferable to others for the encampment of troops, by a very celebrated medical authority. Sir John Pringle confiders the banks of large rivers as extremely proper for this purpofe, on account of a free circulation of air; but I am forry to obferve, that Sir John Pringle's opinion on this occafion appears to have arifen from his theory, rather than that his theory has arifen from obfervation. We have actual experience of the unhealthfulnefs of the muddy banks of large rivers in hot climates; and we have little caufe to dread difeafes, which originate from confined air in America, the Weft-Indies, or perhaps in any country where troops are employed in the field.

I have juft now obferved, that the banks of large rivers, in the opinion of Sir John Pringle, afford the moft eligible fituation with refpect to healthinefs for the encampment of troops. I may add, that the fame author has likewife recommended open grounds for this purpofe, in preference to woods; and that the fame favourite idea, viz. a free circulation of air, . has influenced his advice. I will not contend, that open, dry and cultivated grounds may not be preferable to grounds covered with wood, where the heat of the climate is moderate; but I have no doubt in afferting, that encampments on lands, the woods of which have been newly cut down, as is generally the cafe in times of war, are of all others the moft unhealthful. I have myfelf feen feveral examples of it. Perhaps it is in a great meafure owing to this caufe, that new countries are generally fo fatal to the firft fettlers; as alfo, that troops fuffer fo remarkably in carrying on the fieges of places which are furrounded by woods: it being conftantly obferved, that effluvia from moift lands, when firft expofed to the action of a powerful fun, are always highly pernicious. The Romans, whofe obfervations on fubjects which relate to war, may be oppofed with confidence to the authority of the moft celebrated moderns, were fully fenfible of the advantages of encamping under the fhelter of wood. We learn from Vegetius, that their armies reforted to the cover of a wood, not lefs carefully, than that they avoided the vicinity of fwamps or marfhes. There are in reality various circumftances, which contribute to render fuch fituations both healthy and agreeable. If troops are encamped in the body of a wood, the noxious effluvia, which may be carried by the winds from neighbouring fwamps, are ftopt in their progrefs; the lofty fhade of the trees preferves the air cool and more refrefhing than the atmofphere of the open country; while we know from experience, that moift and fwampy lands do not fend forth their noxious vapours, in any remarkable degree, unlefs where they are acted upon by the heat of a powerful fun.

I fhall only further obferve, with regard to the caufe of intermitting and remitting fevers, that a fpace of time almoft conftantly intervenes between expofure to the noxious effluvia, and the fubfequent appearance of the difeafe. It is not indeed uniformly the fame in all cafes, appearing to depend not only on the concentrated ftate in which the exhalation is applied to the body; but on the general aptitude of the individual, and the various occafional or exciting caufes, which facilitate or refift its operation. It was in a few inftances only, that I faw the difeafe appear before the feventh day. It was oftener the fourteenth, twentieth, or even longer. Upon the whole I may remark, that feptenary periods has a confiderable power in influencing the time of its appearance.

Having offered a few obfervations in the preceding pages, on the diet, exercifes and choice of the quarters or encampments for troops in hot climates; I fhall now add a few hints refpecting medical care and management. It will probably be fuppofed, that no attention with refpect to this fubject has been omitted. Regiments are provided with furgeons, and armies have always been furnifhed with ample hofpital eftablifhments. But this perhaps is not enough. It is neceffary that the duties of thefe ftations be well executed, as well as well defigned. The office of furgeon to a regiment is an office of truft and of primary importance; the appointment to it, however, does not feem in general to be fufficiently attended to. The furgeoncies of regiments, till lately, were allowed to be bought and fold; in confequence of which abufe, little other qualification, came to be required, than the command of the purchafe money. Thus it often happened, that young men, who had attended a courfe of anatomical lectures, or walked the rounds of an hofpital for a few months, came at once to be en rufted with the care of the lives of fix or feven hundred foldiers, who, as they are raifed and maintained at a great expence, deferve, on the fcore of economy, independent of every other confideration, to be well taken care of. It would be fuperfluous to ufe any arguments to prove the prodigality of committing the care of a regiment to men, who have not had profeffional experience in any country, and who are totally unacquainted with the difeafes of the countries to which they are frequently fent. If we are difpofed to believe that there is any thing in medical treatment, we can fcarcely avoid making the conclufion, that many lives are loft from this caufe. It muft not be underftood, that I mean any thing direfpectful to the furgeons of the army, by this infinuation. I know that a regiment is an excellent fchool for medical knowledge; and that the beft practitioners have occafionally appeared in the army; but I wifh ftrongly to inculcate the propriety of obliging candidates for this office, to produce evidence of their qualifications, before they are admitted to fuch an important truft. It is not enough, that a young man, who offers himfelf to take charge of the health of a regiment, fhould know to perform an operation with dexterity. Handling a knife in reality is the leaft part of a regimental furgeon's duty. The office of phyfician is his daily employment, to execute which properly, both years and experience are required. It certainly ought therefore to be an object of concern with thofe who are entrufted with the office of fuperintending the medical appointments of the army, that the candidates. for furgeoncies be obliged to fubmit to fuch trials, as may in fome degree afford proofs of their abilities. It would be a proper regulation, perhaps, that no man be permitted to propofe himfelf for the furgeoncy of a regiment, before he has arrived at fuch an age, as may have furnifhed him with general experience; and further, that he give teftimony of actual abilities by the treatment of difeafes in an hofpital, under the infpection of an able phyfician, to whom the duty will be prefcribed to examine the mode of practice with rigour. A trial of this fort might be better trufted to than the recommendatory letter of a profeffor; or even the diploma of Oxford or Edinburgh. There is not any thing chimerical in the propofal. Nothing in fhort is more practicable; but it is fcarcely to be expected, that men of talents and education will give

themfelves fo much trouble, that they may be admitted into a fervice which holds out few advantages. The
falary of regimental furgeons is fmall; and it is perhaps no paradox to fay, that this is a caufe of great
expence to the nation. The bare fubfiftence, which
the fervice affords, furnifhes no inducement for men
of abilities to enter, at leaft to remain any length of
time in the army, which unfortunately has been confidered in no other light, than as a place, where furgeons may pafs their noviciate; but which they are
generally difpofed to leave, as foon as they are qualified to execute the duty properly. Medical knowledge is gained only by experience but independent
of medical knowledge, an acquaintance with the habits, characters and difpofitions of foldiers is a matter
of fo great importance, that old furgeons, even of inferior abilities as medical men, have generally been
obferved to have a proportionably fmall lift of fick in
their refpective corps. The qualities that are principally required in a regimental furgeon, exclufive of medical learning and knowledge, are acutenefs in difcerning the characters and difpofitions of men, and above
all, boldnefs and decifion in the application of remedies. Life is often loft in unhealthy climates, by
the dilatorinefs and timidity of common practice.

Having mentioned juft now, that there appears to
be a remiffnefs in examining the qualifications that
are requifite for the office of regimental furgeon, I
might perhaps, with equal juftice, extend the remark to the appointments in the general hofpital.—
The power of appointing phyficians or furgeons in
the hofpital, has generally been lodged with the com-
manders in chief, and I might fay, without tranfgref-
fing the bounds of truth, that merit has not always
been the beft claim for promotion. It would be in-
vidious to be more particular in cenfuring what is
paffed. It is only hoped, that the fubject will be en-
quired into, and fuch remedies applied, as will pre-

clude fimilar abufes in future. The general hofpital
has ever been a heavy article in the expences of war;
and if it were fair to form an opinion of the whole,
from the part which I have feen, I fhould not hefi-
tate in declaring, that the eftablifhment is in a great
meafure fuperfluous. I have no doubt, in obtaining
the fuffrages of people of experience, that general
hofpitals are ruinous to military difcipline; that they
promote floth and indolence, the worft difeafe to
which a foldier is liable, and that they extinguifh
very fpeedily the ardour for the fervice of the field.--
There is in fact no exaggeration in the affertion, that
the man, who has fpent two or three months in a general hofpital, is lefs of a foldier than when he was
firft recruited. It is only I may add by habits of exercife, even by toils and fatigue, that men at laft attain the properties of good foldiers: while it is only
by conftant practice of fuch difcipline, that they are
preferved in a ftate fit for the performance of their
duties. Thefe active qualities are fpeedily extinguifhed by the habits of floth and indolence, which
prevail in general hofpitals; but befides this, it is
likewife certain, that cures are often there protracted
to months, which might have been accomplifhed in
the courfe of a few days, if circumftances would have
permitted the men to remain with their regiments.—
Regimental furgeons have many inducements to exert
themfelves in reftoring their men fpeedily to health,
which act only with feeble power on thofe who have
the management of general hofpitals. The former
likewife poffefs fome advantages, of which the latter
are deftitute. They know the habit and difpofitions
of the patient; they fee the difeafe in its firft beginnings, and are enabled to feize the moft favourable
moments for acting with decifion. The above are
confiderations, which ought to make us backward in
removing fick foldiers to general hofpitals; I may
add, that fuch is the nature of military difeafes, that

there does not, perhaps occur one cafe in twenty, which might not be treated properly by the furgeon of the regiment, if attention, and a very little expence were beftowed in providing neceffary accommodation. But befides that, the difeafes of foldiers are feldom of fuch a kind, that they might not be treated properly by regimental furgeons, if government were at the expence of fupplying a few conveniencies.

I may farther obferve, that together with the indolence naturally attached to general hofpitals, and uniformly hurtful to military difcipline, there is often  actual danger to life, by removing men in critical fituations, or by the neceffary intermiffion of medical affiftance, where continual and vigorous exertions are required  The difeafes of hot climates, particularly the fevers of the Weft-Indies, are often moft acute and rapid in their courfe.  The furgeon of a regiment perceives the approach of danger, and, fenfible that his fituation does not enable him to do juftice to his patient, determines to remove him to the general hofpital.  But time is loft before this can be accomplifhed.  It is feldom that any thing is done after it is deemed proper to fend a fick man away; neither does it commonly happen, that any thing material is attempted on the day on which he is received.  Thus one day at leaft, fometimes two are completely loft in cafes, where every moment is of confequence.  Time is precious in the fevers of hot climates; and the decifion or neglect of an hour often determines the fate of a patient.

It is an obfervation, which cannot fail of having frequently occurred to people who have ferved any time in the army, that it would be a very great advantage to the fervice, if fick foldiers could always be taken care of by the refpective furgeons of the regiment.  I have endeavoured to fhew, that the plan is practicable and eafy; and I may further add, that the expence neceffary for fuch an arrangement,

would not amount to one third of what is ufually
fpent in general hofpital eftablifhments. If this idea
were adopted, nothing more would be required, than
that proper lodgings, proper provifions, and a fufficient fupply of medicines, were furnifhed for the
fick ; that the furgeon of the regiment be well qualified for his ftation ; and that an infpector be appointed for a certain portion of troops, to take care
that the duty be well and diligently executed. By
this means a general hofpital, as far as regards medical treatment, might be abolifhed, or at leaft greatly
abridged. Where fighting was expected, extra-fur-
gical affiftance would ftill be neceffary. Such an
infpection of regimental hofpitals, as that I have mentioned, feems to be perfectly fufficient for the care of
the health of an army, in ordinary occafions. The
greateft precautions, however, ought to be taken,
that the infpection does not degenerate into a nominal
duty. The infpector ought to vifit the different
quarters, examine minutely into every article of the
management of the hofpitals, and order that general
reports be publifhed annually ; and that fome mark
of approbation be beftowed upon thofe furgeons, who
appear to have executed their duty with the greateft
diligence and ability. It ought to be a concern of
government, however, that no perfon be appointed
to infpect regimental hofpitals, who is not well acquainted with the difeafes of the climate, in which
the troops happen to ferve.

# N O T E S.

## CHAP. I.

THOUGH I have defcribed the endemic fever
of Jamaica as diftinctly as is in my power;
yet as I have often obferved that we attain more accurate ideas from the detail of a particular hiftory
than from general defcription, I fhall felect from my
notes two or three cafes which may ferve to give a
clearer view of the different fpecies of the difeafe.
And in the firft place I fhall defcribe an inftance of
fever, which was diftinguifhed through the whole of
its courfe by fymptoms of the general inflammatory
diathefis.

(1) Lennox, a foldier of the 60th regiment, aged
40, of a firm and compact habit of body, was feized
on the 3d of December, between eight and nine in
the morning, with a flight horror or fhivering, preceded and accompanied by other ufual marks of fever.
The fymptoms of coldnefs and fhivering went off in
the courfe of eight or ten minutes. A hot fit fucceeded, with a good deal of head-ach, hurried refpiration, confiderable thirft, a ftrong, full, and frequent
pulfe. After a continuance of four or five hours,
fweat began to appear on the head and breaft, which
extending gradually to the extremities brought with
it a tolerable diftinct remiffion of the fever. 2. The
fever appeared to be gone off very completely by
ten o'clock at night. The patient refted well during

the night, and continued in this fame ftate of eafe till about five o'clock in the evening. He then became uneafy and reftlefs, with head-ach and a flight feverifh heat. 5. The feverifh indifpofition declined in the courfe of the night. He became eafier towards morning, and about feven might be faid to be in a ftate of remiffion. About nine a paroxyfin commenced, fimilar to the paroxyfm of the firft day, though with a ftill flighter degree of horror and fhivering; the hot fit ran ftill higher, with much head-ach, thirft, and a ftrong vigorous pulfe. The fweating at laft made its appearance, and the violence of the fever declined: there ftill however remained fome degree of head-ach, pain of the back, and thirft, with an averfion to food, and a more than natural frequency of pulfe. 6. Refted tolerably; but ftill is not free from head-ach and pain of the back: the tongue is dry and foul, and the coat which covers it is fmooth, but of confiderable thicknefs, and of a cream colour. About four in the evening the paroxyfm of a fever made its appearance, fimilar, in fome degree, to the paroxyfm of the fourth, but of a much greater degree of violence. It continued for eight hours, and declined gradually towards morning. 7. There were no perceivable marks of fever at feven in the morning. A little after nine, however, a paroxyfin commenced, fimilar, in every refpect, to the paroxyfm of the fifth. 8. About three in the afternoon a paroxyfm began fimilar to the paroxyfm of the fixth, but ftill more violent. It declined after the ufual duration, and was fucceeded on the ninth by another paroxyfm fimilar to that of the feventh. The remiffion which fucceeded appeared to be ftill more perfect than any of the preceding; the fweat was even more copious, and the pulfe became fofter and more expanded after it than it had hitherto done. 10. A paroxyfm returned about half paft two, fimilar to the paroxyfm of the eighth, but not lefs violent in degree. It terminated, however, in a more fluid and univerfal fweat; the pulfe and the ftate of the fkin returned perfectly to what they were in health; the mucous coat feparated from the tongue; the eye and countenance affumed their natural ferenity, and unequivocal marks of final crifis appeared on the morning of the eleventh. The above cafe is an inftance of the double tertian; the fever of the even day terminated the difeafe; and the pulfe through the whole courfe was vigorous and ftrong, or marks of inflammatory diathefis, in a moderate degree, were conftantly prefent.

(2) Henley, a foldier of the 60th regiment, was feized on the 6th of May, about five in the evening, with a naufea, or unpleafant affection at ftomach, marks of great languor and debility, a flight feeling of coldnefs and horror, a very weak and frequent pulfe, head-ach, pain of the back, and other fymptoms which are ufual in the acceffion of fevers. After a continuance of ten or twelve hours, thefe fymptoms were fo far gone off, that the patient was confidered to be in a ftate of remiffion. 7. The exacerbation of the fever returned again about the fame hour in the evening at which it had firft come on, though without marks of preceding coldnefs or fhivering. The pulfe was fmall, obfcure, and extremely frequent; the heat of the body was not increafed very materially; the thirft was only in a moderate degree, but there was much naufea, an averfion to food, a difpofition to faint in an erect pofture, deep and heavy fighing, tremor of the tongue, and a fad and defponding ftate of the eye and countenance. 8. The fymptoms of fever abated towards morning, and a remiffion, though by no means a diftinct one, took place. The pulfe became fomewhat flower and more expanded; the fighing and anxiety abated a little, and there was evidently a ftate of greater eafe; though there ftill remained marks of great debility, and figns

of fpafmodic ftricture on the furface of the body. The heat was lower than it ufually is in health. About five in the evening the fymptoms, which had prevailed in the former paroxyfms, returned again, but with confiderable aggravation. The head was affected with delirium, and there was a confiderable degree of tremor and ftarting. 9. Eafier in the morning, though the remiffion was in no degree more complete than the former. About the ufual hour in the evening the fame fymptoms returned with aggravation. 10. The remiffion as the former; the heat of the body below natural; the pulfe obfcure and frequent; the figns of debility very great. The exacerbation returned again at the ufual hour; the paroxyfm appeared to be fomewhat more violent; the delirium was higher, the heat greater, and the pulfe acquired rather more ftrength and fulnefs. 11. Eafier in the morning, with a remiffion in every refpect as complete as the former; the pulfe diftinct, and rather more expanded; and, upon the whole, an appearance of rather more vigour. The paroxyfm was renewed in the evening as ufual. 12. Remiffion in the morning rather more complete: more vigour in the pulfe. The exacerbation as ufual. 13. In the morning, inftead of the ufual remiffion, there appeared marks of a complete and final crifis; the fighing, which had been troublefome throughout the courfe of the difeafe, vanifhed; the eye and countenance affumed their ufual ferenity and cheerfulnefs; the pulfe became flower, fofter, and more expanded; and the tongue parted with its coat or covering. The above is an inftance of fever with fymptoms of nervous affection.

(3) Sergeant Negli, on the 2d of November, about eight in the morning, was feized with horror, fhivering, and other Symptoms, which are ufual in the acceffion of fevers. The hot fit did not run very

high, and before evening the paroxyfm was confiderably abated. 3. This patient is now in the ftate of remiffion, the heat of the body is not greater than natural; but the cafe feems to be attended with fome fymptoms which are not very common in the fevers of this country. The countenance is clouded, dark, and grim; the appearance of the eye is fad and defponding; and he expreffes an uneafinefs in his feelings which is not eafily accounted for. 4. The paroxyfm returned about four in the morning. It was greatly more violent than the preceding; and though it might be faid to remit very completely, if we judge by the heat of the body and ftate of the pulfe; yet there ftill remained fome uncommon and unpleafant feelings. The eye and countenance were not only dark and defponding, but the tongue was covered with a flimy mucous coat, through which the red furface appeared obfcurely; there were ftrange and unaccountable twitchings of the ftomach and bowels, difturbed fleep, frightful dreams, and foreboding apprehenfions. 5. A paroxyfm came on this evening near twelve hours fooner than it was expected. After expreffing an eafinefs at ftomach, and throwing up fome matter of a dark colour, he was fuddenly feized with a ftupor and infenfibility, from which he could not be roufed by all the applications of art. He died in about fixteen hours. This cafe affords an inftance of fever with marks of a peculiar malignity. The appearances of danger were fudden and unexpected; and, as it was among the firft inftances of the kind which I had feen, I was difappointed, and in fome degree confounded at the event.

(4) Thomfon a young man aged twenty, after more than ufual exercife in the heat of the fun, was feized with ficknefs, fhiverings, and other figns of fever, about nine o'clock in the morning of the 3d. of February. The pulfe was hard, frequent, and

irritated ; the eye was fad, and fometimes gliftening; the countenance flufhed, but rather dark and over-caft; the refpiration hurried; naufea was troublefome, with a good deal of anxiety and reftleffnefs. The paroxyfm continued long, and did not indeed go off very perfectly at laft. 4. Refted but indifferently; is now fomewhat eafier, though the remiffion is far from being perfect; the thirft is confiderable; the tongue dry and foul, the ftomach loathes all forts of food ; and he feems to be much diftreffed with flatus and ructus ; the ftools are dark-coloured and fœtid ; the pulfe is more frequent than natural, hard and ir-ritated, and the fkin is only partially moift. 5. An exacerbation of fever happened about nine in the morning. The fymptoms were of the fame kind as in the firft paroxyfm, only fomewhat more violent in degree. The anxiety at ftomach was particularly diftreffing, and there appeared ftill more evident marks of putrefcent tendency in the alimentary canal. 6. An uneafy night : an imperfect and obfcure remiffion : the gums redder than they naturally are: the eye has a gliftening appearance, and the countenance is ftill confufed and clouded : the tongue is dry ; the thirft great ; and ructus and flatus are very diftreffing : the pulfe ftill irritated and quick : there is not any very remarkable difpofition to faint in an erect pof-ture : the ftools fœtid. 7. The exacerbation returned about the fame hour as on the fifth, and with ftill greater aggravation : the fymptoms of diftrefs in the ftomach and bowels were particularly alarming ; with naufea, nidorofe belchings, and large watery fœtid ftools. 8. Somewhat eafier in the morning, though the remiffion can only be faid to be obfcure. 9. The exacerbation happened at the fame hour as on the feventh, and continued for nearly the fame length of time. 10. Inftead of obfcure remiffion, marks of final crifis are now evident ; the pulfe is returned nearly to its natural ftate; the eye and

Countenance have affumed their ufual ferenity; the
fkin is moift, and gives no marks of remaining fpafmodic ftricture; the anxiety and ructus have ceafed;
and the ftate of the ftomach and bowels appears to
be almoft natural. The above is an inftance of
fever, in which there were very evident figns of putrefcent tendency in the alimentary canal; even fome
obfcure marks of its progrefs in the general fyftem,
complicated, however, with an irritated ftate of the
vafcular fyftem, or fuch fymptoms as might be confidered as belonging to the apparent inflammatory
diathefis.

(6) Cunningham, a failor, aged twenty-five, was
feized on the 5th of July, about five in the evening,
with ficknefs, fhiverings, head-ach, and the other
ufual figns of the remitting fever of the country. Its
more diftreffing fymptoms were naufea and vomiting.
6. The remiffion is tolerably diftinct; but there is
ftill a good deal of head-ach, thirft, and figns of
debility; the tongue is dry, and the pulfe is more
frequent than natural. The paroxyfm returned
about five in the evening with increafed violence, accompanied with fevere retching, and copious
vomiting of bilious matter. 7. Better in the morning; the vomiting has ceafed, and the remiffion is
tolerably diftinct. The exacerbation returned at
the ufual hour, with the fame diftinguifhing fymptoms of copious bilious difcharges. 8. Remiffion in
the morning as ufual; the exacerbation in the evening as the preceding, with diftreffing and fevere vomiting. 9. The ufual remiffion in the morning.
The paroxyfm likewife recurred in the evening about
the ufual time, but not with the ufual fymptoms.
Inftead of vomiting of bilious matters, there was
fome degree of delirium, tremors, ftartings, and
other fymptoms of nervous affection. 10. Thefe
fymptoms remitted in the morning, but there ftill

remained figns of great irritability and weaknefs.
The fame train of fymptoms returned again in the
evening: the delirium and tremors were ftill in a
higher degree; the pulfe was fmall and frequent;
and there was occafionally a great difpofition to faint
in an erect pofture. 11. Better in the morning,
though there are not yet any marks of crifis. The
exacerbation returned again in the evening, with
fymptoms fimilar to thofe of the preceding paroxyfm.
12. Remiffion in the morning fimilar to the former.
Exacerbation in the evening rather more violent.
13. Remiffion as the former; the pulfe however appears to be rather fuller than it has been fince this
change happened in the circumftances of the difeafe.
The paroxyfm returned at the ufual hour ftill more
violent, though with greater marks of vafcular excitement. 14. Evident marks of crifis: the tongue
begins to part with its covering; the eye and countenance appear clear and animated; the pulfe is flower
and fuller; and the ftate of the fkin does not give
any indication of exifting fpafmodic ftricture. This
cafe prefents an inftance of fever, the firft part of
the courfe of which was diftinguifhed by uncommon bilious difcharges during the time of the paroxyfms; the latter part of it by affection of the nervous fyftem.

# C H A P. II.

1. As I mentioned before that we attain more accurate ideas from the detail of particular cafes than from general histories ; I therefore relate the method of cure, which was purfued in thofe examples which are defcribed in the fixth chapter.

1. Lennox.——On the 4th of December, or fecond day of the difeafe, the folution of falts with a finall portion of emetic tartar was given by a wine glafs full at a time, till it operated plentifully. 5. Some powders of nitre and camphire. 6. Two fcruples of bark were given every two hours during the remiffion, with an injunction that the nitrous powders be repeated during the time of the paroxyfm. 11. The above plan was perfifted in till marks of crifis appeared. Not more than one ounce of bark was given during all the remiffions.

2. *Henley.* 7. The ufual folution of falts was given, but without any addition of emetic tartar. It operated plentifully. 8. The bark was begun this morning, with injunctions that it be adminiftered every two hours during the remiffions. 9. A blifter was applied to the back of the head and neck, with a bolus of camphire, opium, and valerian. Wine was ordered, together with the bark, as foon as the remiffion fhould begin to appear. This plan was perfifted in till the crifis arrived, which was on the 13th.

3. *Negli.* The patient was purged on the 3d with the ufual folution of falts, to which was added fo confiderable a portion of emetic tartar, that it likewife operated by vomit. 4. Bark was given in the

ufual quantity, and at the ufual intervals. 5. As foon
as the fever came on, blifters were applied to the
head, and likewife to the legs; but they produced no
perceivable effect. The patient died, and probably
fell a facrifice to the difeafe, from my not having early
enough perceived the malignity of its nature.

4. *Thompfon.* 4th, The folution of falts with emetic tartar was adminiftered in the prefent cafe as it
had been done in the others. It operated plentifully,
but had no material effect upon the difeafe. 5. Sa-
line draughts in the ftate of effervefcence were given
frequently. Bark and wine were ordered in the remiffions, with as much lemonade as the patient chofe
to drink. 9. Glyfters of cold water, impregnated
with fixed air, were employed two or three times
with apparent benefit. 10. The bark, wine, and faline draughts were given liberally, yet notwithftanding, the difeafe feemed to complete its natural courfe,

5. Cunningham.—6. The naufea and vomit-
ing were fo diftreffing in the firft paroxyfm, that, in
compliance with the patient's earneft entreaties, I
confented to give an emetic. 7. The fymptoms were
aggravated, and the emetic was repeated but without
advantage. 9. Anodynes were given during the paroxyfm with faline draughts in the act of effervefcence. They moderated the vomiting but did not
entirely remove it. Blifters were applied to the head
and legs; bark and wine were given in confiderable
quantity; but the difeafe continued till the 14th without material alteration.

# CHAP. III.

1. As the cold bathing, which I have fo ftrongly recommended in the cure of fevers, has an exterior appearance of being a rafh and hazardous remedy, I fhall relate fome cafes which may enable the reader to judge more precifely of its real effects. Cold-bathing I may remark, appears to have been occafionally employed by the Greek and Roman phyficians, after the time of the Emperor Auguftus; but I was only a young man when I went out to the Weft-Indies, and cannot pretend to fay that I was acquainted with the writings of thofe phyficians, or that I poffeffed much knowledge of difeafes, except the little that could be retained from a curfory hearing of univerfity lectures. The firft hints of this practice were therefore accidental, and arofe from a converfation I had with the mafter of the veffel, in which I went paffenger. This perfon commanded a tranfport in the war 1756, and was prefent at the fiege of the Havannah. As he was talking one day of the ftate of the fleet, he mentioned accidently, that fome men were fent aboard of his fhip ill of fevers; feveral of whom, he obferved, jumped into the fea during the dilirium which attended the paroxyfms of the difeafe. Some of them, as might be expected, were drowned; but the moft part of thofe who were recovered from the waves appeared to be greatly benefited by the ducking. The fact, which, from the veracity of the man, I thought I could depend upon, ftruck me ftrongly, and I refolved, in my own mind, to bring it to the teft of experiment as foon as an opportunity fhould offer: neither was it long after my arrival in Jamaica, that I had occafion to vifit a failor whofe fituation feemed to juftify fuch a trial. The poor

man was aboard of a fhip, which lay at anchor about
a mile from the fhore. He had been ill two days ;
the delirium ran high; his eyes were red and inflamed; his refpiration was hurried; he was anxious and
reftlefs in a high degree, whilft together with thofe
marks of excitement, he was occafionally languid and
difpofed to faint. His fkin being dirty furnifhed an
oftenfible excufe for trying this remedy. But it was
previoufly thought proper to draw fome blood from
the arm; which being done, fome buckets of falt
water were dafhed on the fhoulders. He was now
laid in bed: a copious fweat enfued, fucceeded by a
diftinct remiffion, and a total change in the nature of
the fymptoms. The fuccefs I met with in this inftance was more than I had expected ; I was therefore encouraged to try the fame mode of bathing in
a perfon who came under my care fome weeks there-
after, and who had been ill of a fever fix or feven
days. This patient had been bled and bliftered;—
emetics and cathartics had been likewife employed,
and bark had been given in the ufual manner, for the
three laft days. The fever, however, had now in a
great meafure loft its type. The man was low and
languid; his eyes were dim; his vifion indiftinct ;—
his pulfe was fmall and frequent, and, when the head
was raifed from the pillow, not to be felt. Though
it did not appear that he could reafonably be expected
to live long, I ftill wifhed to get him conveyed to the
deck, that a trial might be made of the effects of cold
bathing ; but the fituation was fo ticklifh, that I felt
fome uneafinefs in fetting about it. At laft he was
lifted through the hatch-way in a blanket, though I
muft confefs that I was not without apprehenfions
that he might die under my hands. Some wine was
then poured down his throat; and he was fprinkled
with cold falt water, as he lay on the deck. Appearing to be fomewhat invigorated by this procefs,
he was raifed up very gently, and feveral buckets of

the fea-water were dafhed about his head and fhoulders. He was then laid in bed; the pulfe foon became large and full. I left him in a copious fweat, and was agreeably furprifed next day to find him fitting on the deck, to which he had walked on his own feet. I fhall only mention another inftance of the good effects of cold bathing in the fevers of the Weft-Indies, which is perhaps more decifive than either of the former. A boy, aged fourteen, had been ill of a fever feven or eight days. Nothing had been omitted, in point of treatment, which is ufual to be done in fimilar cafes. Bark and wine had been carried as far as could be ferviceable, or even fafe; yet death feemed to be approaching faft. The fuccefs of cold bathing, in fome inftances fimilar to the prefent, fo far exceeded my expectation that I was induced to make trial of it in the cafe before me, though I was not altogether without apprehenfions that death might be the confequence of the attempt. The bufinefs, however, was accomplifhed without accident; and next day the boy was able, not only to fit up in bed, but even to walk over the floor. After inftances fo unequivocal as the above, it would be fuperfluous to mention any others. I fhall only add, that I have tried the remedy, in various fituations, always with fafety, generally with aftonifhing fuccefs; fo that I cannot forbear recommending it even at an early period, in the fevers of the Weft-Indies. It communicates tone and vigour to the powers of life, and diminifhes irritability in a degree far fuperior to all other cordials or fedatives. The bathing was managed in the following manner: the water, which was required to be of a refrefhing degree of coolnefs, was generally dafhed by means of a bucket on the head and fhoulders. It was likewife found that its good effects were heightened, in fome cafes, by previous bleeding, and by the previous ufe of warm bathing. This may feem a rafh practice to thofe who argue

C c

without experience; but, fetting afide the authority
of the ancients, we find it confirmed by the example
of a perfon who was not a phyfician, and who, therefore may be fuppofed to be lefs under the influence of
a favourite opinion from which he might be led to
difguife the truth. Bufbequius, who was fent on an
embaffy to Soliman the Great, was obliged to travel
to Amafia, where the Sultan then fojourned. In his
return home he was feized with a continued fever,
and very feverely harraffed by it. The difeafe gained
fo much ground during the journey, that he found it
neceffary to ftop at Conftantinople to attend to the
recovery of his health. The practice which was
adopted to effect this may appear to be fingular, and
by many, perhaps, will be thought to be hazardous
and rafh. He mentions, that, after enjoying the
luxury of warm bathing, he was fuddenly fprinkled
with cold water. His words are, " Idem, fcilicet,
Quaquelbenus me a balneo exeuntem frigida perfun-
debat, quæ res etfi erat molefta, tamen magnopere
juvabat." ( Iter. Conftant. p. ) His phyfician, Quaquelbenus, who feems to have been a man of excellent judgment and careful obfervation, had probably
learnt the practice in his travels in Afia, as it does
not appear to have been commonly known in Europe
at that time.

# C H A P. IV.

1. But besides these testimonies of physicians, in favour of the practice of drenching with cold water, the memoirs of Baron Trenck furnish us with a curious and very convincing proof of the efficacy of this remedy, in extinguishing, almost like a charm, the violence of a burning fever. The Baron, when ill of a fever in the prison at Madgeborough, unfortunately broke the pitcher which contained his daily allowance of water. The fever was violent, and he suffered the most inexpressible torments of thirst, for the space of twenty-four hours. When the usual supply was brought to him next day, he seized the pitcher with eagerness, and drank the water with such avidity, and in so great quantity as is scarcely credible. The consequence was a total removal of the disease. To this I might add an instance, which happened to myself at Savanna in Georgia, in the y ar 1779. In the excessive hot weather of the month of July, I was seized with the endemic of the country, in a more violent degree than was commonly seen. In the third paroxysm of the disease, my desire for cold water was ravenous. A pitcher of it was drawn from the pump, which I drank off instantly without the least abatement of the thirst. The draught was repeated in a few minutes, in quantity not less than a quart. The thirst was effectually quenched, and the fever seemed to vanish. But though the fever appeared to be extinguished as it were by a charm ; yet the stomach and hypochondria became distended, yellowness of the eye and countenance succeeded, with a considerable degree of debility which remained for two or three days. I must, however, remark with regard to this case, that the effects were not the same as they have been usually reported to be by

authors. The fever was extinguilhed; but neither vomiting, fweat, or any other fenfible evacuation enfued. The ancients, I may further obferve, feem to have adminiftered cold drink only in the advanced ftate of fever, when figns of coction began to appear; in which cafe, it is impoffible to form a certain opinion of its precife fuccefs. That cold water may be employed with effect, it is neceffary that the thirft be intenfe, perhaps that it be purpofely provoked, and that it be fully and completely fatiated. If managed in this manner, it probably will not often fail of extinguifhing the fever; yet I muft not omit to mention, that unlefs it is managed with a great deal of caution and judgment, it may alfo often irrecoverably extinguifh the powers of life.

2. In fupport of this opinion, I fhall mention a cafe, which fell under my own obfervation about a twelvemonth ago. I was called to a young man, a failor, ill of a fever of a very dangerous and alarming kind. It was the eighth day of the difcafe before I faw him. He had not been hitherto in the leaft benefited by any thing that was tried; neither did any remedy which I could think of, though employed with almoft defperate boldnefs, in any degree check the progrefs of the difeafe. The power of fpeech was loft, and even fwallowing was performed with difficulty; the eye was languid, nay almoft without motion; the countenance was ghaftly; and many livid fpots, fome of them nearly the fize of a fix-pence, made their appearance on different parts of the body. I propofed bathing, and the friends of the young man, confidering the fituation as defperate, confented that I fhould make a trial of it; more, perhaps, to comply with my defire, than from expectation of any benefit that might refult from it. But in fetting about it, it unfortunately happened, that an utenfil proper for the purpofe could not be procured, fo that

we were compelled to be contented with a general fomentation. This was applied in as complete a manner as circumſtances would permit. A blanket was ſoaked in warm ſalt water, and the body was wrapped in it from head to foot. In a ſhort time the ſkin became ſoft and warm, ſweat began to flow ; the eye and countenance began to reſume their animation, which had been almoſt extinguiſhed, the pulſe roſe ; ſwallowing was performed with leſs difficulty ; and next day the colour of the ſpots was evidently brighter. So far the change was favourable ; but a regular ſupply of wine and cordials having been neglected during the following night, the pulſe ſunk, and things returned nearly to their former ſituation. The fomentation was again repeated, in conſequence of which the extremities and ſurface of the ſkin became warm and moiſt, an effect which was no ſooner produced, than the blanket was ſuddenly removed, and the face and breaſt, particularly, were ſprinkled with cold water, in which a large portion of ſalt was diſſolved. The cold had the effect to cauſe the patient to ſhrink at the firſt, yet in a ſhort time he appeared to be refreſhed very remarkably. The powers of life grew gradually ſtronger ; though the marks of criſis were not very evident for ſeveral days. To the above I might add ſome other inſtances, where effects were ſimilar ; but I avoid ſwelling the notes to too great extent, by entering into particular details. I ſhall therefore only obſerve in general, that cold bathing was uſually of ſervice. It imparted general tone and vigour to the powers of life, and by increaſing the activity of the vaſcular ſyſtem, probably ſometimes rendered the criſis complete, where it naturally would not have been ſo ; but I cannot venture to ſay that I ever carried it ſo far that the diſeaſe could be ſaid to be precipitately extinguiſhed by it.

# NOTES

## TO THE

## CHAPTER UPON YELLOW FEVER.

---

(*a*) IN compliance with the language of medical authors, I have defcribed the following difeafe under the name of Yellow Fever, though I am per- fectly fenfible, that the appellation is not by any means proper. There are fome inftances of the difeafe perhaps where yellownefs does not at all appear, and in no one does it ordinarily fhew itfelf till the latter ftages. I know alfo that moft of the practi- tioners of Jamaica confider it only as an aggravated fpecies of the remittent ; the common endemic of hot climates. It appeared to me I muft confefs in a dif- ferent light ; but I fhall attempt to defcribe the two difeafes accurately, and leave it to the reader to judge for himfelf. It may not however be improper in this place to take notice of the opinion of Dr. Mofeley. Dr. Mofeley has lately written a treatife on this difeafe, and endeavoured to perfuade us that it is no other than the Καυσος, or ardent fever of the ancients. But the yellow fever of the Weft-Indies, by Dr. Mofeley's own confeffion, is in fome manner peculiar to ftrangers newly arrived in tropical climates. The Καυσος we are informed, made its appearance in the iflands of the Archipelago, and on the coafts of the contiguous continents indifcriminately among men and women, natives or foreigners : in fact it has not, as far as I can perceive, any claim to be confidered as a diftinct fpecies of difeafe.  If I rightly underftand the fpirit of Hippocrates,  or the defcription of the ftill more accurate Arctæus, Καυσος in reality is only an acci-dental condition of the common endemic of the coun-try, where the force of the fever is chiefly exerted upon the ftomach and alimentary canal.  In this manner it appears frequently in Jamaica, and in th· fouthern provinces of America.  In the hot months of fummer, it appears occafionally in every climate : and is not neceffarily accompanied with, nor does it depend upon a general inflammatory diathefis of the fyftem for its exiftence.

6. Authors feem generally to have attributed the black colour of the vomitings obferved in this difeafe to blood effufed into the cavity of the ftomach ; but the falfity of this opinion is fufficiently proved by the appearances which are obferved on diffection.

T A E  E N D.

# B O O K S

PRINTED FOR, AND SOLD BY

## R O B E R T  C A M P B E L L,

*No.* 54, *South Second-Street.*

---

## A SYSTEM OF SURGERY.

By Benjamin Bell, Member of the Royal Colleges of
furgeons of Ireland and Edinburgh, one of the
Surgeons to the Royal Infirmary, and Fellow of
the Royal Society of Edinburgh, in four vols.
illuftrated with 100 copperplates. Price 4 dol-
lars 50 cents.

---

## A TREATISE ON THE THEORY AND MANAGEMENT OF ULCERS,

With a diftertation on White Swellings of the
Joints, to which is prefixed an Eftay on the Chi-
rugical Treatment of Inflammation, and its Confequences. By Benjamin Bell, F.R.S. Ed. &c. &c.
Price 1 dol. 50 cents.

---

## FIRST LINES OF THE PRACTICE OF PHYSIC.

By William Cullen M.D. late Profeffor of the
practice of Phyfic in the Univerfity of Edinburgh,
&c. &c. in 2 volumes; with practical and ex-
planatory notes, by John Rotherham, M.D. To
which is prefixed the Life of the Author. Price
3 dols. 75 cents.

---

## THE ART OF PREVENTING DISEASES AND RESTORING HEALTH,

Founded on rational principles and adapted to perfons of every capacity. By George Wallis, M.D.
S. M. S. editor of the laft edition of Motherby's
Medical Dictionary, and Sydenham's Works with
notes, &c. &c. Price 2 dols.

## A TREATISE

On the management of Female Complaints and of Children in early infancy. By Alexander Hamilton. M. D. &c. &c. Price 1 dol.

## THE LIFE OF BARON FREDERICK TRENCK.

Containing his adventures, his cruel and exceffive fufferings during ten years imprifonment at the fortrefs of Magdeburg, by command of the late King of Pruffia; alfo anecdotes, hiftorical, political, and perfonal. To which is now added his adventures in France. 1 dol.

## THE SEASONS.

James Thomfon. Price 75 cents.

## THE MISCELLANEOUS WORKS OF Dr. GOLDSMITH;

Containing his effays and poems. Price 75 cents.

## MENTORIA, OR THE YOUNG LADY's FRIEND.

By Mrs. Rowfon, of the New Theatre, Philadelphia, author of the Inquifitor, Fille de Chambre, Victoria, Charlotte, &c. &c. Price 75 cents.

## A SIMPLE STORY.

By Mrs. Inchbald. Price 1 dol.

## THE NATURAL HISTORY OF THE BIBLE,

Or a defcription of all the beafts, birds, fifhes, infects, reptiles, trees, plants, metals, precious ftones, &c. mentioned in the Sacred Scriptures, collected from the beft authorities, and alphabetically arranged. By Thaddeus M. Harris, librarian of Harvard univerfity, Cambridge. Price 87 cents.

## THE LADIES LITERARY COMPANION,

Or a collection of effays for the inftruction and amufement of the female fex. Price 50 cents.

# THE PHILOSOPHY OF NATURAL HIS-TORY.
By William Smellie, member of the Antiquarian and Royal Society of Edinburgh.  Price 2 dols.

# THE SORROWS OF WERTER.
A German tale.  Price 62 cents.

# THE BEAUTIES OF HISTORY.
Two volumes.  Price 2 dols.

# A TREATISE CONCERNING RELIGIOUS AFFECTIONS.
In three parts.  Part I.——Concerning the nature of the affections, and their importance in religion. Part II.——Shewing what are no certain figns that religious affections are gracious, or that they are not.  Part III.——Shewing what are diftinguifhing figns of truly gracious and holy affections.  By Jonathan Edwards, A. M. and paftor of the firft church in Northampton.  Price 1 dol.

# HISTORY OF REDEMPTION.
On a plan entirely original : exhibiting the gradual difcovery and accomplifhment of the divine pur-pofe in the falvation of man : including a compre-henfive view of church hiftory, and the fulfilment of fcripture prophecies.  By the late Rev. Jona-than Edwards, prefident of the college of New-Jerfey.  To which are now added, notes, hiftori-cal, critical, and theological; with the life and ex-perience of the author.  Price 2 dols.

# A SERMON
On the freedom and happinefs of the United States of America, preached in Carlifle, on the 5th of October 1794, and publifhed at the requeft of the officers of the Philadelphia and Lancafter troops of light horfe.  By Robert Davidfon, D. D. paftor of the Prefbyterian church in Carlifle, and one of the profeffors in Dickinfon college.  Price 20 cents·

## THE MILLENNIUM;

Or, the thoufand years of profperity promifed to the church of God in the old teftament and in the new; fhortly to commence, and to be carried on to perfection. Price 1 dol. 25 cents.

## SERMONS,

By Hugh Blair, D. D. F. R. S. Edinburgh, one of the minifters of the high church, and profeffor of rhetoric and belles-lettres, in the univerfity of Edinburgh, 2 vols. Price 2 dols.
Same book, vol. 3rd, being the fame as the 4th volume of the Britifh edition. Price 1 dol.

## BEAUTIES OF HERVEY;

Or, defcriptive, picturefque, and inftructive paffages, felected from the works of this defervedly admired author, viz. Meditations among the tombs—reflections on a flower garden—defcant on the creation—contemplations on the night—the ftarry-heavens, and a winter piece—the moft important, interefting, and picturefque paffages from Theron and Afpafio—letters and fermons—mifcellaneous tracts—religious education of daughters, and remarks on lord Bolingbroke's letters. To which are added—Memoirs of the author's life and character, with an elegiac poem on his death. Price 80 cents.

## WATTS's PSALMS AND HYMNS,

Bound together and feparate; bibles, teftaments, fpelling-books, primers, &c. &c. all of which will be fold by wholefale and retail, on the very loweft terms.

R. CAMPBELL, has in the prefs and fpeedily will be publifhed, Bell on the Venereal Difeafe : in one vol. 8vo. Price, neatly bound, 2 dols.